T0267400

NANOBIOTECHNOLOGY
BASIC AND APPLIED ASPECTS

NANOBIOTECHNOLOGY
Basic and Applied Aspects

Edited by

Arunava Goswami

and

Samrat Roy Choudhury

UNION BRIDGE PRESS

UNION BRIDGE BOOKS
An imprint of Wimbledon Publishing Company Limited (WPC)
First published in the United Kingdom in 2017 by
Union Bridge Books
This edition first published in UK and USA 2017
UNION BRIDGE BOOKS
75–76 Blackfriars Road
London SE1 8HA

www.unionbridgebooks.com

British Library Cataloguing-in-Publication Data
A catalogue record for this book is available from the British Library.

Library of Congress Cataloging-in-Publication Data
A catalog record for this book has been requested.

ISBN-13: 978-1-78308-737-2 (Hbk)
ISBN-10: 1-78308-737-4 (Hbk)

This title is also available as an e-book.

CONTENTS

PREFACE

In the 1980s, nanotechnology offered the potential fulfillment of a dream for anyone concerned with the future of the planet. It was becoming apparent that we would one day need to reduce the quantity of materials and energy used in making all our machines. Nanoscience and nanotechnology are going to provide the breakthroughs for turning the possibilities into certainties, and that is why they are at the forefront of modern research. The fast growing economy in this area requires experts who have an outstanding knowledge of nanoscience in combination with the skills to apply this knowledge in new products. These newborn scientific disciplines are situated at the interface between physics, chemistry, materials science, microelectronics, biochemistry, and biotechnology. Control of these disciplines therefore requires an academic and multidisciplinary scientific education.

Also, the Government of India, in May 2007, has approved the launch of a Mission on Nano Science and Technology (Nano Mission) with an allocation of Rs. 1000 crore for 5 years. Capacity-building in this upcoming area of research will be of utmost importance for the Nano Mission so that India emerges as a global knowledge-hub in this field. For this, research on fundamental aspects of Nano Science and training of large number of manpower will receive prime attention. Equally importantly, the Nano Mission will strive for development of products and processes for national development, especially in areas of national relevance like safe drinking water, materials development, sensors development, drug delivery, etc. All this means that nanotechnology and nanoscience will be taught far more widely in the coming days.

This book is a small endeavor on our part to provide young students an introduction to these disciplines, the new tools that are being developed, the immense possibilities and the challenges that are likely to be faced.

BOOK SYNOPSIS

The book is a comprehensive look at how nanoscience can be applied to biological systems. The authors have used extensively, work carried out in their laboratories to serve as examples for illustrating the ideas they outline here. This hopefully will provide a ringside view of nanoscience in applications of immediate concern to humanity. The first chapter introduces the reader to the idea of nanoscience, a brief history of the idea and new possibilities in the field. The second chapter gives a detailed account of the various methods by which nanoparticles can be synthesized, which essentially includes chemical synthesis (employing both organic and inorganic materials) and biological synthesis (employing plants or microbes). The third chapter describes the structural and functional attributes of specific non-metallic nanoparticles namely, carbon nanotubes, selenium and sulfur nanoparticles. In addition, an overview of the currently developed methods, physicochemical characterizations and biological implications of the synthesized particles is provided. The next chapter gives a bird's eye view of nanoscale materials being investigated for biomedical uses. Magnetic nanoparticles (MNPs) have gained significant attention due to their intrinsic magnetic properties, which enable tracking through radiology, magnetic resonance (MR) imaging etc. This class of NPs includes metallic, bimetallic, and iron oxide magnetic nanoparticles (MNPs). The latter has been widely favored because of their inoffensive toxicity profile and reactive surface that can be readily modified with biocompatible coatings as well as targeting, imaging, and therapeutic properties. This flexibility has led to MNP use in magnetic separation, biosensor, in vivo medical imaging, drug delivery, tissue repair, and hyperthermia applications .The sixth chapter deals with polymer nanocomposites. These are microscopically heterophase, contain embedded isotropic/anisotropic particulate entities of nanometric size within soft polymer matrices. These nanosized entities offer unmatched physico-mechanicals, chemical and rheological properties which not only excel in ordinary applications but also have shown great promise towards many advanced applications in the near future. The last chapter informs the reader about natural nanoparticles, their widespread occurrence in nature, both as geological and biological entities. The book also includes a description of the special technological devices like TEM, SEM, DLS, AFM and their deployment for studying nanoparticles.

Chapter 1

AN INTRODUCTION TO THE SCOPE OF NANOSCIENCE AND NANOTECHNOLOGY

Indrani Roy and Arunava Goswami

The Greek word for dwarf is "Nano." The first mention of some of the distinguishing concepts in nanotechnology (but predating use of that name) was in 1867 by James Clerk Maxwell, when he proposed a tiny entity known as Maxwell's Demon, able to handle individual molecules, as a thought experiment. The first observations and size measurements of nano-particles were made during the first decade of the twentieth century. They are mostly associated with Richard Adolf Zsigmondy, who made a detailed study of gold sols and other nanomaterials, with sizes as low as 10 nm and less. He published a book in 1914, and used an ultramicroscope that employs the *dark field* method for seeing particles with sizes much less than light wavelength. Zsigmondy was also the first to use the term **nanometer** explicitly for characterizing particle size. He determined it as 1/1,000,000 of a millimeter. He developed the first system classification based on particle size in the nanometer range.

More significant developments in characterizing nanomaterials took place in the second half of the twentieth century. The prefix "nano" was officially recognized in 1956. The first use of the concepts found in "nanotechnology" was in a talk given by physicist Richard Feynman on December 29 1959. Feynman described a process by which the ability to manipulate individual atoms appeared plausible.

The term "nanotechnology" was defined by Tokyo Science University professor Norio Tanuguchi in a 1974 paper as follows: "'Nano-technology' mainly consists of the processing of, separation, consolidation, and deformation of materials by one atom or by one molecule." In the 1980s, the basic idea of this definition was explored in greater depth by Dr K. Eric Drexler, who promoted the technological significance of nano-scale phenomena and devices through speeches and the books *Engines of Creation: The Coming Era of Nanotechnology* (1986) and *Nanosystems: Molecular Machinery, Manufacturing, and Computation*, after which the term acquired its current sense. *Engines of Creation* is considered the first book on the topic of nanotechnology. Nanotechnology and nanoscience got started in the early 1980s with two major developments: the birth of cluster science and the invention of the scanning tunneling microscope (STM). This development led to the discovery of fullerenes in 1985 and carbon nanotubes a few years later. In another development, the synthesis and properties of semiconductor nanocrystals was studied. This led to a fast increasing number of metal and metal oxide nanoparticles and quantum dots. The atomic force microscope (AFM or SFM) was invented six years after the STM, and made possible the visualization of nanomaterials (Fig. 1). Don Eigler moved the first individual atom twenty years ago, and shortly afterward he wrote IBM's name with 35 Xenon atoms (Fig. 2). Moving atoms had big consequences in that it made the idea of assembling devices atom by atom very real. IBM has built on that nanotechnology foundation, storing information on specific gold atoms, collecting carbon monoxide molecules into computer logic circuits, and pursuing a vision for vastly more compact computing technology.

The following illustration, titled "The Scale of Things," created by the US Department of Energy, provides a comparison of various objects to help you begin to envision exactly how small a nanometer is.

The chart starts with objects that can be seen by the unaided eye, such as an ant, and progresses to objects about a nanometer or less in size, such as the ATP molecule used in all living systems to store energy from food (Fig. 3).

Now that you have an idea of how small a scale nanotechnologists work with, consider the challenge they face. Such an image helps you imagine the problem scientists have while working with nanoparticles.

Sometimes a distinction is made between nanotechnology and nanoscience, the latter focusing on the observation and study of phenomena at the nanometer scale. The distinction is not of great importance. Nanoscience suggests a solid body of theory upon which a technology can be built. Such theory is still inchoate, however, and the nanotechnologist is as likely to contribute to it as the nanoscientist. It is best to use the term "nanotechnology" in an all-embracing sense (Fig. 4, Fig. 5).

Although nanoparticles are generally considered an invention of modern science, they actually have a very long history. Humans have unwittingly employed nanotechnology for thousands of years, for example in making steel, paintings and in vulcanizing rubber. The best known surviving example is the Lycurgus Cup from the Roman Empire, fourth century AD. The cup glows red or green depending on whether light is transmitted or reflected from the cup. The effect was achieved by mixing minute amounts of silver and gold in the glass. Pottery from the Middle Ages and Renaissance often retains a distinct gold or copper-colored metallic luster. This so-called luster is caused by a metallic film applied to the transparent surface of a glazing. The luster can still be visible if the film has resisted atmospheric oxidation and other weathering. The luster originates within the film itself, which contains silver and copper nanoparticles dispersed homogeneously in the glassy matrix of the ceramic glaze. These nanoparticles were created by artisans by mixing copper and silver salts and oxides together with vinegar, ochre and clay on the surface of previously-glazed pottery. The object was then placed into a kiln and heated to about 600°C in a reducing atmosphere. In the heat, the glaze would soften, causing the copper and silver ions to migrate into the outer layers of the glaze. There, the reducing atmosphere reduces the ions back to metals, which then come together forming the nanoparticles that give the color and the optical effects (Fig. 6).

Luster technique showed that ancient craftsmen had a rather sophisticated empirical knowledge of materials. Carbon black is the most famous example of a nanoparticulate material that has been produced in quantity for millennia. It was used to make "Wootz" steel, high-grade steel that was highly prized and much sought-after across several regions of the world for over nearly two millennia. Wootz steel is characterized by a pattern of bands or sheets of micro carbides. It was developed in India around 300 BC. There is archaeological evidence of the manu-facturing process in South India from that time. Wootz steel was widely exported and traded throughout ancient Europe and the Arab world. It became particularly famous in the Middle East, where it was known as Damascus steel. Studies have found existence of carbon nanoparticles in the famous sword of Tipu Sultan and the Ajanta Paintings (Fig. 7, Fig. 8).

70% of carbon black now produced is used as a pigment and in the reinforcing phase in manufacturing automobile tires. Carbon black also helps conduct heat away from the tread and belt area of the tire, reducing thermal damage and increasing tire life. Carbon black particles are also employed in some radar-absorbent materials and in photocopier and laser printer toner.

Michael Faraday provided the first description, in scientific terms, of the optical properties of nanometer-scale metals in his classic 1857 paper [Faraday Michael (1857). "Experimental relations of gold (and other metals) to light". Phil. Trans. Roy. Soc. London **147**: 145–181.] In a subsequent paper, Turner [Turner, T. (1908). "Transparent Silver and Other Metallic Films".*Proc. Roy. Soc. Lond. A* **81** (548): 301–310] points out that:

> *"It is well known that when thin leaves of gold or silver are mounted upon glass and heated to a temperature which is well below a red heat (~500°C), a remarkable change of properties takes place, whereby the continuity of the metallic film is destroyed. The result is that white light is now freely transmitted; and reflection is correspondingly diminished, while the electrical resistivity is enormously increased."*

The transition from microparticles to nanoparticles can lead to a number of changes in physical properties. Two of the major factors in this are the increase in the ratio of surface-area-to-volume, and the size of the particle moving into the realm where quantum effects predominate. The increase in the surface-area-to-volume-ratio, which is a gradual progression as the particle gets smaller, leads to an increasing dominance of the behavior of atoms on the surface of a particle over that of those in the interior of the particle. This affects both the properties of the particle in isolation and its interaction with other molecules. Nanoparticles are of great scientific interest as they are effectively a bridge between bulk materials and atomic or molecular structures. A bulk material should have constant physical properties, regardless of its size. But at the nano-scale, size-dependent properties are often observed. Thus the properties of materials change as their size approaches the nanoscale and as the percentage of atoms at the surface of a material becomes significant. For bulk materials larger than one micrometer (or micron), the percentage of atoms at the surface is insignificant in relation to the number of atoms in the bulk of the material. The interesting and sometimes unexpected properties of nanoparticles are therefore largely due to the large surface area of the material, which dominates over the contributions made by the small bulk of the material. Nanoparticles often possess unexpected optical properties as they are small enough to confine their electrons and produce quantum effects. For example, gold nanoparticles appear deep red to black in solution. Nanoparticles of usually yellow gold and gray silicon appear red in color. Gold nanoparticles melt at much lower temperatures (~300°C for 2.5 nm size) than the gold slabs

(1064°C). Absorption of solar radiation in photovoltaic cells is much higher in materials composed of nanoparticles than it is in thin films of continuous sheets of material, i.e. the smaller the particles, the greater the solar absorption. Other size-dependent property changes include quantum confinement in semiconductor particles, surface plasmon resonance in some metal particles and superparamagnetism in magnetic materials. Ironically, the changes in physical properties are not always desirable. Ferromagnetic materials smaller than 10 nm can switch their magnetization direction using room temperature thermal energy, thus making them unsuitable for memory storage. Suspensions of nanoparticles are possible since the interaction of the particle surface with the solvent is strong enough to overcome density differences, which otherwise usually result in a material either sinking or floating in a liquid. High surface area is a critical factor in the performance of such technologies as fuel cells and batteries. The huge surface area of nanoparticles also results in a lot of interactions between the intermixed materials in nanocomposites, leading to special properties such as increased strength and/or increased chemical/heat resistance. The fact that nanoparticles have dimensions below the critical wavelength of light renders them transparent. This transparency makes them useful for applications in packaging, cosmetics and coatings.

Some of the properties of nanoparticles may not be predicted simply by understanding the increasing influence of surface atoms or quantum effects. It has been recently shown that perfectly formed silicon nanospheres, with diameters of between 40 and 100 nm, are not just harder than silicon but among the hardest materials known, falling between sapphire and diamond.

What is New in Nanotechnology?

Nanoscience has been present all along in the traditional disciplines of Chemistry, Biology and Physics. Chemistry is a powerful contender for claiming nanotechnology under its domain. Chemistry deals with the manipulation of molecules and hence is familiar with nanometer dimensions. However the chemist does not control systems in the way the engineer does. Molecules mostly reside in their free energy minima and it requires special ingenuity, intuition and luck to steer their precursors along the paths to the desired end product.

Physics provides the answers to the questions raised by the special properties of matter at nano scale. Size is mostly a relative term, but quantum mechanics offers a definition of absolute smallness: a system is absolutely small if it is perturbed by the act of observing it. Thus a photon is usually destroyed by the act of observation, or its state is irretrievably altered. Most nanosystems are not small enough for this to be the case. Nevertheless, quantum effects are needed to understand certain nano-objects, for example, the small clusters of atoms called quantum dots, nanodots or nanoparticles. These objects are tiny spheres of a solid, typically a semiconductor. In condensed matter, electrons are no longer the point particles that they are believed to be in free space, but have extension, quantified by their Bohr radius, which can vary from a few to hundreds of nanometers, depending on the material. It is possible to make nanoparticles smaller than the Bohr radius of electrons in them. Thus the electrons are subjected to quantum confinement, with the observable effect that the optical absorption and fluorescent emission of the particle are shifted towards higher energies, the magnitude of the shift depending on the particle size. Additionally, a number of physical (mechanical, electrical, optical, etc.) properties change when compared to macroscopic systems. One example is the increase in surface-area-to-volume ratio altering mechanical, thermal and catalytic properties of materials. Quantum effects must also be considered with ultra-miniaturized electron circuitry-single electron devices. Much of the fascination with nanotechnology stems from these quantum and surface phenomena that matter exhibits at the nanoscale.

Biology is considered to provide living proof of the principle of nanotechnology-nanomechanical devices, nanoreactors, nanosensors and nanoassemblers. Biological structures at macromolecular and supramolecular scales are apparently assembled using the principles of self-assembly. These structures, mostly protein based, often combine extraordinary lightness with extraordinary strength. Two amazing examples are the F_1ATPase enzyme that uses a proton gradient across the cell membrane in which it is embedded to synthesise ATP, and the Type III Secretion System (TTSS) — a spherical assembly of needles found on the surface of certain pathogenic bacteria and used to inject toxins into their targets.

Physics, chemistry and biology strongly overlap with nanoscience, but differ essentially from nanotechnology, which seeks to impose control over materials and devices at this scale. Nanoscience and nanotechnology could therefore be defined via the convergence of chemistry, biology, physics and engineering.

Key Elements of Nanotechnology

The technology of realization can be conveniently divided into **fabrication** and **metrology**. While fabrication is dealt with here, metrology is discussed in the next chapter.

There is a wide variety of techniques for fabricating nanoparticles. These essentially fall into three categories. The first two are "bottom-up" approaches while the third is a "top-down" method of making nanoparticles.

Condensation from a vapor: This method is used to make metallic and metal ceramic nanoparticles. It involves evaporation of a solid metal followed by rapid condensation to form nanosized clusters that settle in the form of a powder. Various approaches to vaporizing the metal can be used and variation of the medium into which the vapor is released affects the nature and size of the particles. Inert gases are used to avoid oxidation when creating metal nanoparticles, whereas a reactive oxygen atmosphere is used to produce metal oxide ceramic nanoparticles. Final particle size is controlled by variation of parameters such as temperature, gas environment and evaporation rate.

Chemical synthesis: This consists of growing nanoparticles in a liquid medium composed of various reactants. This is typified by the sol-gel approach and is also used to create quantum dots. Chemical techniques are generally better than vapor condensation techniques for controlling the final shape of the particles. The ultimate size of the nanoparticles might be dictated, as with vapor condensation approaches, by stopping the process when the desired size is reached, by choosing chemicals that are stable and stop growing at a certain size. The approaches are generally low-cost and high-volume, but contaminations from the precursor chemicals can be a problem. This can interfere with one of the common uses of nanoparticles — sintering — to create surface coatings.

Solid state processes: Grinding or milling can be used to create nanoparticles. The milling material, milling time and atmospheric medium affect resultant nanoparticle size and properties. The approach can be used to produce nanoparticles from materials that do not readily lend themselves to the two previous techniques. Contamination from milling material can be an issue.

Possibilities of Nanotechnology

The Application of Nanotechnology to Energy Production

Here are some interesting methods being explored of using nanotechnology to produce more efficient and cost-effective energy:

An epoxy containing carbon nanotubes is being used to make windmill blades. Stronger and lower weight blades are made possible by the use of nanotube-filled epoxy. The resulting longer blades increase the amount of electricity generated by each windmill.

Researchers have used sheets of nanotubes to build thermocells that generate electricity when the sides of the cell are at different temperatures. These nanotube sheets could be wrapped around hot pipes, such as the exhaust pipe from a car, to generate electricity from heat that is usually wasted. They have prepared graphene layers to increase the binding energy of hydrogen to the graphene surface in a fuel tank, resulting in a higher amount of hydrogen storage and therefore a lighter weight fuel tank. This may help in the development of practical hydrogen-fueled cars.

Researchers have developed piezoelectric nanofibers that are flexible enough to be woven into clothing. The fibers can turn normal motion into electricity to power cell phones and other mobile electronic devices. Lubricants using inorganic buckyballs show significantly reduced friction (Fig. 9).

Companies have developed nanotech solar cells that can be manufactured at significantly lower cost than conventional solar cells. They are currently developing batteries using nanomaterials. One such battery will be as good as new after sitting on the shelf for decades. Another battery can be recharged significantly faster than conventional batteries. Nanotechnology is being used to reduce the cost of catalysts used in fuel cells. These catalysts produce hydrogen ions from fuel such as methanol. It is also being used to improve the efficiency of membranes used in fuel cells to separate hydrogen ions from other gases, such as like oxygen.

Nanotechnology can address the shortage of fossil fuels, such as diesel and gasoline, by making the production of fuels from low grade raw materials economical.

Nanotechnology in Agriculture and Food Science

The current global population is more than seven billion, with fifty per cent living in Asia. A large proportion of those living in developing countries face daily food shortages as a result of environmental impacts or political instability. For developing countries, the drive is to develop drought and pest-resistant crops, which also maximize yield. In developed countries, the food industry is driven by consumer demand, which is currently for fresher and healthier foodstuffs. This is a big business. For example, the food industry in the UK is booming with an annual growth rate of 5.2 per cent and the demand for fresh food has increased by 10 per cent in the last few years. The application of nanotechnology to the agricultural and food industries was first addressed by a United States Department of Agriculture road map published in September 2003. The prediction is that nanotechnology will transform the entire food industry, changing the way food is produced, processed, packaged, transported, and consumed:

Nanoparticles are being developed that will deliver vitamins or other nutrients in food and beverages without affecting the taste or appearance. These nanoparticles actually encapsulate the nutrients and carry them through the stomach into the bloodstream.

Researchers are using silicate nanoparticles to provide a barrier to gasses (for example oxygen), or moisture, in a plastic film used for packaging. This could reduce the possibly of food spoiling or drying out.

Zinc oxide nanoparticles can be incorporated into plastic packaging to block UV rays and provide anti-bacterial protection, while improving the strength and stability of the plastic film.

Nanosensors are being developed that can detect bacteria and other contaminates, such as salmonella, at a packaging plant. This will allow for frequent testing at a much lower cost than sending samples to a lab for analysis. This point-of-packaging testing, if conducted properly, has the potential to dramatically reduce the chance of contaminated food reaching grocery store shelves (Fig. 10).

Research is also being conducted to develop nanocapsules containing nutrients that would be released when nanosensors detect a vitamin deficiency in your body. Basically this research could result in a super vitamin storage system in the human body that delivers the nutrients when needed.

"Interactive" foods are being developed that would allow you to choose the desired flavor and color. Nanocapsules that contain flavor or color enhancers are embedded in the food, inert until a hungry consumer triggers them. The method has not yet been published, so it will be interesting to see how this particular trick is accomplished.

Researchers are also working on pesticides encapsulated in nanoparticles that only release pesticide within an insect's stomach, minimizing the contamination of plants themselves.

Another development being pursued is a network of nanosensors and dispensers used throughout a farm field. The sensors recognize when a plant needs nutrients or water, before there is any sign that the plant is deficient. The dispensers then release fertilizer, nutrients, or water as needed, optimizing the growth of each plant in the field, one by one.

Nanotechnology in Medicine

Nanotechnology in medicine involves applications of nanoparticles currently under development, as well as longer range research involving the use of manufactured nano-robots to make repairs at the cellular level (sometimes referred to as "nanomedicine").

For example, nanoparticles that deliver chemotherapy drugs directly to cancer cells are under development. Tests are in progress for targeted delivery of chemotherapy drugs. Final approval for their use with cancer patients is pending. Oral administration of drugs that currently are delivered by injection may be possible in many cases. The drug is encapsulated in a nanoparticle which helps it pass through the stomach to deliver the drug into the bloodstream. There are efforts underway to develop oral administration of several different (multiple) drugs using a variety of nanoparticles.

Buckyballs may be used to trap free radicals generated during an allergic reaction and block the resulting inflammation . Nanoshells may be used to concentrate the heat from infrared light to destroy cancer cells with minimal damage to surrounding healthy cells. Nanospectra biosciences has developed such a treatment using nanoshells illuminated by an infrared laser that has been approved for a pilot trial with human patients. This is intended to be used in place radiation therapy with much less damage to healthy tissue (Fig. 11).

Quantum Dots (q dots) may be used in the future for locating (cancerous) tumors in patients, and in the near term for performing diagnostic tests in samples. Iron oxide nanoparticles can used to improve MRI (magnetic resonance imaging) of cancerous tumors. The nanoparticles are coated with a peptide that binds to a cancerous tumor. Once the nanoparticles are attached to the tumor, the magnetic property of the iron oxide enhances the images from the Magnetic Resonance Imaging scan.

Aluminosilicate nanoparticles can more quickly reduce bleeding in trauma patients by absorbing water, causing blood in a wound to clot quickly by converting prothrombin to thrombin.

Nanofibers can stimulate the production of cartilage in damaged joints. Nanoparticles may be used, when inhaled, to stimulate an immune response to fight respiratory viruses.Nanoparticles can attach to proteins or other molecules, allowing detection of disease indicators in a lab sample at a very early stage. There are several efforts to develop nanoparticle disease detection systems underway. One system uses gold nanoparticles, while another system uses magnetic nanoparticles to identify specimens, including proteins, nucleic acids, and other materials.

One of the earliest nanomedicine applications was the use of nanocrystalline silver, which is as an antimicrobial agent for the treatment of wounds. A nanoparticle cream has been shown to fight staphylococcus infections. Studies on mice have shown that using the nanoparticle cream to release nitric oxide gas at the site of staphylococcus abscesses significantly reduces the infection.

Burn dressing that is coated with nanocapsules containing antibiotics is being developed. If an infection starts, the harmful bacteria in the wound cause the nanocapsules to break open, releasing the antibiotics. This allows much quicker treatment of an infection and reduces the number of times a dressing has to be changed. A welcome idea in the early study stages is the elimination of bacterial infections in a patient within minutes, instead of delivering treatment with antibiotics over a period of weeks.

Nanotechnology in Electronics (Nanoelectronics)

Nanoelectronics hold some answers for how we might increase the capabilities of electronics devices while we reduce their weight and power consumption. Some of the nanoelectronics areas under development are:

Improvement of display screens on electronics devices, which involves reducing power consumption while decreasing the weight and thickness of the screens. Building transistors from carbon nanotubes to enable minimum transistor dimensions of a few nanometers and developing techniques to manufacture integrated circuits built with nanotube transistors. Using electrodes made from nanowires that would enable flat panel displays to be flexible as well as thinner than current flat panel displays.

Use of Micro Electro Mechanical Systems (MEMS) techniques to control an array of probes whose tips have a radius of a few nanometers. These probes are used to write and read data onto a polymer film, with the aim of producing memory chips with a density of one terabyte per square inch or greater (Fig. 12, Fig. 13).

Some more uses are:

Building transistors with single-atom-thick graphene film to enable very high speed functioning, and combining gold nanoparticles with organic molecules to create a transistor known as a NOMFET (Nanoparticle Organic Memory Field-Effect Transistor).

Use of carbon nanotubes to direct electrons to illuminate pixels, resulting in a lightweight, millimeter thick "nanoemmissive" display panel; making integrated circuits with features that can be measured in nanometers (nm), such as the process that allows the production of integrated circuits with 45 nm-wide transistor gates. Nanosized magnetic rings to make Magnetoresistive Random Access Memory (MRAM) may allow memory density of 400 GB per square inch. Development of molecular-sized transistors which may allow us to shrink the width of transistor gates to approximately one nm which will significantly increase transistor density in integrated circuits (Fig. 14).

One very successful attempt at bringing the future into the present is the Cyborg insect developed by DARPA(Defense Advanced Research Projects Agency) of United States Department of Defense.

The HI-MEMS (Hybrid Insect Micro Electrmechanical Systems) program is aimed at developing tightly coupled machine-insect interfaces by placing micro-mechanical systems inside the insects during the early stages of metamorphosis. These early stages include the caterpillar and the pupae stages. Since a majority of the tissue development in insects occurs in the later stages of metamorphosis, the renewed tissue growth around the MEMS will tend to heal, and form a reliable and stable tissue-machine interface. The goal of the MEMS, inside the insects, will be to control the locomotion by obtaining motion trajectories either from GPS coordinates, or using RF, optical, ultrasonic signals based remote control. HI-MEMS – insects with various sorts of different embedded MEMS sensors (like video cameras, audio microphones and chemical sniffers)could penetrate enemy territory in swarms. The HI-MEMS swarms could then perform reconnaissance missions beyond the capabilities of bulky human soldiers (Fig. 15).

Nanoparticles in the Atmosphere
Temporally and spatially, atmospheric nanoparticles are highly variable in number, concentration and composition. Many are liquid droplets or semi-volatile materials, but solid or low-volatility nanoparticles are also widespread, and are those most amenable to study using techniques familiar to geoscientists. Many atmospheric nanoparticles are anthropogenic, although emissions from trees and other plants dominate in some regions, and particles from sea spray dominate elsewhere. From time to time, volcanoes also deliver large quantities of nanoparticles into the atmosphere. The concentrations of nanoparticles are greatly affected by environmental conditions and depend strongly on emission intensities, proximity to sources and meteorological conditions.

In general, the highest number concentrations occur in urban areas, and these nanoparticles also have the greatest effect on human health. Theoretically, they are also those over which we have greatest control in terms of amelioration. In rural areas, natural sources dominate, although anthropogenic sources can be significant (Seinfeld and Pandis, 2006). In order to place nanoparticles into context across the wide size range of particles in the atmosphere, it is useful to note the large differences among number, volume (and mass), and surface area as a function of size. The most continuous and intimate contact the average person has with nanoparticles is almost surely through the air, which is replete with them. Nanoparticles are being generated continuously and in large numbers by vehicles and industries in urban areas and by vegetation and sea spray in rural areas. Volcanoes are sporadic sources of huge numbers. Nanoparticles have large surface-area-to-volume ratios and react rapidly in the atmosphere, commonly growing into particles large enough to interact with radiation and to have serious consequences for visibility and local, regional and global climate. They also have potentially significant health effects.

Challenges Posed by a New Technology

As is the case with most emerging areas of risk, nanotechnology challenges us with many unknowns. These challenges are further complicated by the fact that few risk-related forecasts have been scientifically confirmed. Many industries are extremely optimistic about the opportunities associated with nanotechnology.

Public response to this new technology, as well as the legal climate, will depend upon how much accurate information is available. We believe that managing the unknowns associated with the development and use of nanotechnology will not be much different from gauging the risks involved with environmental liability (EL) or employee practices liability (EPL).

Tools To Study Nanomaterials, Both Natural And Synthetic

The ability to measure and characterize materials (determine their size, shape and physical properties) at the nanoscale is vital if nanomaterials and devices are to be produced to a high degree of accuracy and reliability, and if the applications of nanotechnologies are to be realized.

The human eye has a resolution of 0.2 mm and an optical microscope has a resolving power of 0.2 μm (based on the visible light). We are thus unable to "see" anything at the atomic, nanometer scale. The technology for viewing objects is keeping pace with the manufacture of nanoscale devices. Devices used to view nanoparticles are based on the principle that nanoparticles are not governed by classic mechanics (Fig. 16).

This section is devoted to some such instruments.

Dynamic light scattering (also known as **Photon Correlation Spectroscopy** or **Quasi-Elastic Light Scattering**) is a technique in physics which can be used to determine the size-distribution profile of small particles in suspension or polymers in solution, and is often the first tool used to check whether a material under consideration has particles in the nanodomain. It can also be used to probe the behavior of complex fluids such as concentrated polymer solutions.

Brownian Motion DLS measures Brownian motion and relates this to the size of the particles. Brownian motion is the random movement of particles due to bombardment by the solvent molecules that surround them. Normally DLS is concerned with the measurement of particles suspended within a liquid.

The larger the particle, the slower the Brownian motion will be. Smaller particles are "kicked" further by the solvent molecules and move more rapidly. An accurately known temperature is necessary for DLS because knowledge of the viscosity is required (the viscosity of a liquid is related to its temperature). The temperature also needs to be stable, otherwise convection currents in the sample will cause non-random movements that will ruin the correct interpretation of size. The velocity of the Brownian motion is defined by a property known as the translational diffusion coefficient (usually given the symbol "D," the Hydrodynamic Diameter). The size of a particle is calculated from the translational diffusion coefficient by using the Stokes Einstein equation:

d H kTD = ( )3πη, where:

d (H) = hydrodynamic diameter

D = translational diffusion coefficient

k = Boltzmann's constant

T = absolute temperature

η = viscosity

Note that the diameter that is measured in DLS is a value that refers to how a particle diffuses within a fluid, so it is referred to as a hydrodynamic diameter. The diameter that is obtained by this technique is the diameter of a sphere that has the same translational diffusion coefficient as the particle. The translational diffusion coefficient will depend not only on the size of the particle "core," but also on any surface structure, as well as the concentration and type of ions in the medium.

Rayleigh Scattering

If the particles are small compared to the wavelength of the laser used (typically less than $d = \lambda/10$ or around 60 nm for a He-Ne laser), then the scattering from a particle illuminated by a vertically polarized laser will be essentially isotropic, i.e. equal in all directions.

The Rayleigh approximation tells us that $I \propto d^6$ and also that $I \propto 1/\lambda^4$, where I = intensity of light scattered, d = particle diameter and λ = laser wavelength. The d^6 term tells us that a 50 nm particle will scatter 10^6 or one million times as much light as a 5 nm particle. Hence there is a danger that the light from the larger particles will swamp the scattered light from the smaller ones. This d^6 factor also means it is difficult with DLS to measure, say, a mixture of 1000 nm and 10 nm particles because the contribution to the total light scattered by the small particles will be extremely small. The inverse relationship to λ^4 means that a higher scattering intensity is obtained as the wavelength of the laser used decreases.

A typical dynamic light scattering system comprises six main components. Firstly, a laser (1) provides a light source to illuminate the sample contained in a cell (2). For dilute concentrations, most of the laser beam passes through the sample, but some is scattered by the particles within the sample at all angles. A detector (3) is used to measure the scattered light. The intensity of scattered light must be within a specific range for the detector to successfully measure it. If too much light is detected, the detector will become saturated. To overcome this, an attenuator (4) is used to reduce the intensity of the laser source and hence reduce the intensity of scattering. For samples that do not scatter much light, such as very small particles or samples of low concentration, the amount of scattered light must be increased. In this situation, the attenuator will allow more laser light through to the sample. For samples that scatter more light, such as large particles or samples at higher concentration, the intensity of scattered light must be decreased. The scattering intensity signal from the detector is passed to a digital processing board called a correlator (5). The correlator compares the scattering intensity at successive time intervals to derive the rate at which the intensity is varying. This correlator information is then passed to a computer (6), where the software will analyze the data and derive size information.

Transmission Electron Microscopy (TEM) is microscopy technique whereby a beam of electrons is transmitted through an ultra-thin specimen, interacting with the specimen as it passes through. An image is formed from the interaction of the electrons transmitted through the specimen; the image is magnified and focused onto an imaging device, such as a fluorescent screen, by a sensor such as a CCD camera.

TEMs are capable of imaging at a significantly higher resolution than light microscopes, owing to the small de Broglie wavelength of electrons. This enables the instrument's user to examine fine detail — even as small as a single column of atoms, which is tens of thousands times smaller than the smallest resolvable object in a light microscope. TEM forms a major analysis method in a range of scientific fields, in both physical and biological sciences. TEMs find application in cancer research, virology, materials science as well as pollution and semiconductor research.

At smaller magnifications, TEM image contrast is due to absorption of electrons in the material, due to the thickness and composition of the material. At higher magnifications, complex wave interactions modulate the intensity of the image, requiring expert analysis of observed images. Alternate modes of use allow for the TEM to observe modulations in chemical identity, crystal orientation, electronic structure and sample-induced electron phase shift as well as the regular absorption-based imaging.

Sample preparation in TEM can be a complex procedure. TEM specimens are required to be at most hundreds of nanometers thick, as, unlike neutron or X-Ray radiation, the electron beam interacts readily with the sample — an effect that increases roughly with atomic number squared (z^2). High quality samples will have a thickness that is comparable to the mean free path of the electrons that travel through the samples, which may be only a few tens of nanometers. Preparation of TEM specimens is specific to the material under analysis and the desired information to be obtained from the specimen. As such, many generic techniques have been used for the preparation of the required thin sections.

Materials that have dimensions small enough to be electron transparent, such as powders or nanotubes, can be quickly prepared by the deposition of a dilute sample containing the specimen onto support grids or films. In the biological sciences, in order to withstand the instrument vacuum and facilitate handling, biological specimens can be fixated using either a negative staining material such as uranyl acetate or by plastic embedding. Alternatively, samples may be held at liquid nitrogen temperatures after embedding in vitreous ice. In material science and metallurgy the specimens tend to be naturally resistant to vacuum, but still must be prepared as a thin foil or etched, so some portion of the specimen is thin enough for the beam to penetrate. Constraints on the thickness of the material may be limited by the scattering cross-section of the atoms from which the material is comprised.

There are a number of drawbacks to the TEM technique. Many materials require extensive sample preparation to produce a sample thin enough to be electron transparent, which makes TEM analysis a relatively time-consuming process with a low throughput of samples. The structure of the sample may also be changed during the preparation process. The field of view is also relatively small, raising the possibility that the region analyzed may not be characteristic of the whole sample. There is potential that the sample may be damaged by the electron beam, particularly in the case of biological materials (Fig. 17).

Unlike the TEM, where electrons of the high voltage beam carry the image of the specimen, the electron beam of the **Scanning Electron Microscope** (SEM) does not at any time carry a complete image of the specimen. The SEM produces images by probing the specimen with a focused electron beam that is scanned across a rectangular area of the specimen (raster scanning). At each point on the specimen the incident electron beam loses some energy, and that lost energy is converted into other forms, such as heat, emission of low-energy secondary electrons, light emission (cathodoluminescence) or x-ray emission. The display of the SEM maps the varying intensity of any of these signals into the image in a position corresponding to the position of the beam on the specimen when the signal was generated. The image is constructed from signals produced by a secondary electron detector, the normal or conventional imaging mode in most SEMs.

Generally, the image resolution of an SEM is about an order of magnitude poorer than that of a TEM. However, because the SEM image relies on surface processes rather than transmission, it is able to image bulk samples up to many centimeters in size and, depending on instrument design and settings, has a great depth of field, and so can produce images that are good representations of the three-dimensional shape of the sample (Fig. 18, Fig. 19).

Reflection Electron Microscope (REM)

In the **Reflection Electron Microscope** (REM), as in the TEM, an electron beam is incident on a surface, but instead of using the transmission (TEM) or secondary electrons (SEM), the reflected beam of elastically scattered electrons is detected. This technique is typically coupled with **Reflection High-Energy Electron Diffraction** (RHEED) and **Reflection High-Energy Loss Spectrum** (RHELS). Another variation is **Spin-Polarized Low-Energy Electron Microscopy** (SPLEEM), which is used for looking at the microstructure of magnetic domains.

Scanning Transmission Electron Microscope (STEM): The STEM rasters a focused incident probe across a specimen that (as with the TEM) has been thinned to facilitate detection of electrons scattered *through* the specimen. The high resolution of the TEM is thus possible in STEM. The focusing action (and aberrations) occurs before the electrons hit the specimen in the STEM, but afterward in the TEM. The STEMs use of SEM-like beam-rastering simplifies annular dark-field imaging and other analytical techniques, but also means that image data is acquired in serial rather than in parallel fashion.

Low-Voltage Electron Microscope (LVEM): The Low-Voltage Electron Microscope (LVEM) is a combination of SEM, TEM and STEM in one instrument, which operates at relatively low electron-accelerating voltage of 5 kV. Low voltage increases image contrast, which is especially important for biological specimens. This increase in contrast significantly reduces or even eliminates the need to stain. Sectioned samples generally need to be thinner than they would be for conventional TEM (20-65 nm). Resolutions of a few nm are possible in TEM, SEM and STEM modes.

Electron microscopes are expensive to build and maintain, but the capital and running costs of confocal light microscope systems now overlap with those of basic electron microscopes. They are dynamic rather than static in their operation, requiring extremely stable high-voltage supplies, extremely stable currents to each electromagnetic coil/lens, continuously-pumped high- or ultra-high-vacuum systems, and a cooling water supply circulation through the lenses and pumps. As they are very sensitive to vibration and external magnetic fields, microscopes designed to achieve high resolutions must be housed in stable buildings (sometimes underground) with special services such as magnetic field cancelling systems. Some desktop low-voltage electron microscopes have TEM capabilities at very low voltages (around 5 kV) without stringent voltage supply, lens coil current, cooling water or vibration isolation requirements and as such are much less expensive to buy and far easier to install and maintain, but do not have the same ultra-high (atomic scale) resolution capabilities as the larger instruments.

The samples largely have to be viewed in vacuum, as the molecules that make up air would scatter the electrons. One exception is the environmental scanning electron microscope, which allows hydrated samples to be viewed in a low-pressure (up to 20 Torr/2.7 kPa), wet environment.

Scanning electron microscopes usually image conductive or semi-conductive materials best. Non-conductive materials can be imaged by an environmental scanning electron microscope. A common preparation technique is to coat the sample with a several-nanometer layer of conductive material, such as gold, from a sputtering machine; however, this process has the potential to disturb delicate samples.

Small, stable specimens such as carbon nanotubes, diatom frustules and small mineral crystals (asbestos fibers, for example) require no special treatment before being examined in the electron microscope. Samples of hydrated materials, including almost all biological specimens, have to be prepared in various ways to stabilize them, reduce their thickness (ultrathin sectioning) and increase their electron optical contrast (staining).

The **Environmental Scanning Electron Microscope (ESEM)** is a scanning electron microscope (SEM) that allows for the option of collecting electron micrographs of specimens that are "wet," uncoated, or both, by allowing for a gaseous environment in the specimen chamber (Fig. 20).

The ESEM, with its specialized electron detectors, differential pumping systems to allow for the transfer of the electron beam from the high vacuums in the gun area to the high pressures attainable in its specimen chamber, make it a complete and unique instrument designed for the purpose of imaging specimens in their natural state.

The ESEM deviates substantially from a SEM in several respects. The presence of gas around a specimen creates new possibilities unique to ESEM:

1. Gas ionization in the sample chamber eliminates the charging artifacts, typically seen with nonconductive samples. So the specimens do not need to be coated with a conductive film. ESEM gets rid of the preparation process.

2. The ESEM can image wet, dirty and oily samples. The contaminants do not damage or degrade the image quality.

3. ESEM can acquire electron images from samples as hot as 1500°C, because the Environmental Secondary Detector (ESD) is insensitive to heat.

4. The detector is also insensitive to light. Light from the sample, for example incandescence from heated samples, cathodoluminescence and fluorescence do not interfere with imaging.

5. ESEM eliminates the need for conductive coating, so delicate structure, which was often damaged during the sample preparation, can be imaged.

6. ESEM can acquire x-ray data from insulating samples at high accelerating voltage.

7. Eliminating the need for sample preparation, particularly the need for conductive coating, makes it possible to investigate specimen in dynamic processes, such as tension, compression, deformation, crack propagation, adhesion, heating, cooling, freezing, melting, hydration, dehydration and sublimation.

As a result, specimens can be examined faster and more easily, avoiding complex and time-consuming preparation methods without modifying the natural surface or creating artifacts by the preceding preparation work or the vacuum of the SEM. Gas/liquid/solid interactions can be studied dynamically in situ and in real time, or recorded for post-processing. Temperature variations from subzero to above 1000°C and various ancillary devices for specimen micro-manipulation have become a new reality. Biological specimens can be maintained fresh and live. Therefore, ESEM constitutes a radical breakthrough from conventional electron microscopy, for which the vacuum condition precluded the advantages of electron beam imaging becoming universal. The main disadvantage arises from the limitation of the distance in the specimen chamber over which the electron beam remains usable in the gaseous environment. The useful distance of the specimen from the PLA1 (Pressure Limiting Aperture) is a function of accelerating voltage, beam current, nature and pressure of gas, and of the aperture diameter used. This distance varies from around 10 mm to a fraction of a millimeter as the gas pressure may vary from low vacuum to one atmosphere. Furthermore, as the pressure can be lowered to a very low level, the ESEM will revert to a typical SEM operation without the above disadvantage.

Concomitant with the limitation of useful specimen distance is the minimum magnification possible, since, at very high pressure, the distance becomes so small that the field of view is limited by the PLA1size. In the very low magnification range of SEM, overlapping with the upper magnification of a light microscope, the superior field is limited to a varying degree by the ESEM mode. The degree of this limitation strongly depends on instrument design.

As X-rays are also generated by the surrounding gas and also come from a larger specimen area than in SEM, special algorithms are required to deduct the effects of gas on the information extracted during analysis.

The presence of gas may yield unwanted effects in certain applications, but the extent of these will only become clear as further research and development is undertaken to minimize and control radiation effects.

No commercial instrument is as yet available that is in conformity with all the principles of an optimum design, so any further limitations listed are characteristic of the existing instruments and not of the ESEM technique, in general.

Field Emission Scanning Microscope (FESM): Nowadays, three-dimensional features can be observed due to the large depth of field available in the FESEM. The addition of energy dispersive X-ray detector combined with digital image processing is a powerful tool in the study of materials, allowing good chemical analysis of the material. The FESEM is a major tool in materials science research and development.

Atomic Force Microscopy (AFM): The Atomic Force Microscope was developed to overcome a basic drawback with STM — that it can only image conducting or semiconducting surfaces. The AFM, however, has the advantage of imaging almost any type of surface, including polymers, ceramics, composites, glass and biological samples.

Binnig, Quate, and Gerber invented the Atomic Force Microscope in 1985. Their original AFM consisted of a diamond shard attached to a strip of gold foil. The diamond tip contacted the surface directly, with the interatomic van der Waals forces providing the interaction mechanism. Detection of the cantilever's vertical movement was done with a second tip — an STM placed above the cantilever.

AFM Probe Deflection

Today, most AFMs use a laser beam deflection system, introduced by Meyer and Amer, in which a laser is reflected from the back of the reflective AFM lever and onto a position-sensitive detector. AFM tips and cantilevers are micro-fabricated from Si or Si_3N_4. Typical tip radius is from a few to 10s of nm.

Measuring Forces

Because the atomic force microscope relies on the forces between the tip and sample, knowing these forces is important for proper imaging. The force is not measured directly but calculated by measuring the deflection of the lever, and knowing the stiffness of the cantilever. Hook's law gives $F = -kz$, where F is the force, k is the stiffness of the lever, and z is the distance the lever is bent.

AFM Modes of Operation: Because of AFM's versatility, it has been applied to a large number of research topics. The Atomic Force Microscope has also gone through many modifications for specific application requirements.

Contact Mode: The first and foremost mode of operation, contact mode, is widely used. As the tip is raster-scanned across the surface, it is deflected as it moves over the surface corrugation. In constant force mode, the tip is constantly adjusted to maintain a constant deflection, and therefore constant height above the surface. It is this adjustment that is displayed as data. However, the ability to track the surface in this manner is limited by the feedback circuit. Sometimes the tip is allowed to scan without this adjustment, and one measures only the deflection. This is useful for small, high-speed atomic resolution scans, and is known as variable-deflection mode.

Because the tip is in hard contact with the surface, the stiffness of the lever needs to be less than the effective spring constantly holding atoms together, which is of the order of 1–10 nN/nm. Most contact mode levers have a spring constant of < 1N/m.

Lateral Force Microscopy: LFM measures frictional forces on a surface. By measuring the "twist" of the cantilever, rather than merely its deflection, one can qualitatively determine areas of higher and lower friction.

Noncontact Mode: Noncontact mode belongs to a family of AC modes, which refers to the use of an oscillating cantilever. A stiff cantilever is oscillated in the attractive regime, meaning that the tip is quite close to the sample, but not touching it (hence, "noncontact"). The forces between the tip and sample are quite low, of the order of pN (10^{-12} N). The detection scheme is based on measuring changes to the resonant frequency or amplitude of the cantilever.

Tapping Mode: This is also is also referred to as intermittent-contact or the more general term, Dynamic Force Mode (DFM).

A stiff cantilever is oscillated closer to the sample than in noncontact mode. Part of the oscillation extends into the repulsive regime, so the tip intermittently touches or "taps" the surface. Very stiff cantilevers are typically used, as tips can get "stuck" in the water contamination layer.

The advantage of tapping the surface is improved lateral resolution on soft samples. Lateral forces such as drag, common in contact mode, are virtually eliminated. For poorly adsorbed specimens on a substrate surface, the advantage is clearly seen.

Force modulation: Force modulation refers to a method used to probe properties of materials through sample/tip interactions. The tip (or sample) is oscillated at a high frequency and pushed into the repulsive regime. The slope of the force-distance curve is measured which is correlated to the sample's elasticity. The data can be acquired along with topography, which allows comparison of both height and material properties.

Phase Imaging: In Phase mode imaging, the phase shift of the oscillating cantilever relative to the driving signal is measured. This phase shift can be correlated with specific material properties that effect the tip/sample interaction. The phase shift can be used to differentiate areas on a sample with such differing properties as friction, adhesion, and viscoelasticity. The techniques are used simultaneously with DFM mode, so topography can be measured as well.

X-ray scattering techniques: These are a family of non-destructive analytical techniques which reveal information about the crystallographic structure, chemical composition, and physical properties of materials and thin films. These techniques are based on observing the scattered intensity of an X-ray beam hitting a sample as a function of incident and scattered angle, polarization and wavelength or energy. X-ray diffraction yields the atomic structure of materials and is based on the elastic scattering of X-rays from the electron clouds of the individual atoms in the system. The most comprehensive description of scattering from crystals is given by the dynamical theory of diffraction.

- Single-crystal X-ray diffraction is a technique used to solve the complete structure of crystalline materials, ranging from simple inorganic solids to complex macromolecules, such as proteins.

- Powder diffraction (XRD) is a technique used to characterize the crystallographic structure, crystallite size (grain size), and preferred orientation in polycrystalline or powdered solid samples. Powder diffraction is commonly used to identify unknown substances by comparing diffraction data against a database maintained by the International Centre for Diffraction Data. Powder diffraction is also a common method for determining strains in crystalline materials.

- Thin film diffraction and grazing incidence X-ray diffraction may be used to characterize the crystallographic structure and preferred orientation of substrate-anchored thin films.

- High-resolution X-ray diffraction is used to characterize thickness, crystallographic structure and strain in thin epitaxial films. It employs parallel-beam optics.

- X-ray pole figure analysis enables one to analyze and determine the distribution of crystalline orientations within a crystalline thin-film sample.

- X-ray rocking curve analysis is used to quantify grain size and mosaic spread in crystalline materials.

References

1. Biswas, N., Rahman, A., Datta, A., Goswami, A., Brahmachary R.L.(2010) Nanoparticle surface as activation site, Journal of Nanoscience and Nanotechnology, 10: 1–5.

2. Drexler, Eric K. (1981) Molecular engineering: An approach to the development of general capabilities for molecular manipulation *Proc. Natl.Acad.Sci.USA* Vol.78 (9):5275-5278

3. Drexler, Eric K. (1986) "Engines of Creation 2.0." New York: Anchor Books

4. Feynman, Richard. (1960) "There is Plenty of Room at the Bottom." Caltech Engineering and Science, Volume 23(5): 22-3

5. Goswami, A., Roy, I., Sengupta, S., Debnath, N. (2010) Novel applications of solid and liquid formulations of nanoparticles against insect pests and pathogens. Thin Solid Films, 519: 1252–1257

6. Majumder D. D., Banerjee R., Ulrichs Ch., Mewis I. and Goswami A. (2007) Nano-materials: Science of bottom-up and top-down. IETE Journal of Research, 24(1): 9-23. Review and Results.

7. Majumder, D.D., Ulrichs, C.H., Mewis, I., Weishaupt, B., Majumder, D., Ghosh, A., Thakur, A.R., Brahmachary, R. L., Banerjee, R., Rahman, A.,

Debnath, N., Seth, D., Das, S., Roy, I., Sagar ,P., Schulz, C., Linh, N.Q. and Goswami, A. (2007): "Current Status and Future Trends of Nanoscale Technology and Its Impact on Modern Computing, Biology, Medicine and Agricultural Biotechnology." (At the International Conference on Computing: Theory and Applications, ICCTA, March 5–7, 2007, India). Conference Publication proceedings, IEEE Press (2007):563–72.

8. Mao, Chengde (2004): "The Emergence of Complexity: Lessons from DNA." PLoS Biology2 (12):2036–2038.

9. Seeman, Nadrian C. (June 2004). "Nanotechnology and the double helix". Scientific American 290 (6): 64–75.

10. Seinfeld, John H. and Pandis, Spyros N. (2006). "Atmospheric Chemistry and Physics: From Air pollution to Climate Change," 2nd edition. John Wiley & Sons, NJ USA.

11. Turner, T. (1908). "Transparent Silver and Other Metallic Films", Proceeding of the Royal. Society London 81 (548): 301–310

12. Rahman, A., Seth, D., Mukhopadhyaya, S.K., Brahmachary, R.L., Ulrichs, Ch. and Goswami, A. (2009) Surface functionalized amorphous nanosilica and microsilica with nanopores as promising tools in biomedicine. Naturwissenschaften, 96:31–38.

13. Ramsden J. J. (2005) What is nanotechnology? Nanotechnology Perceptions1:3–17

14. Salata, O.V. (2004) "Applications of Nanoparticles in Biology and Medicine." In Journal of Nanobiotechnology, 2:3

Electronic references

1. Vaughan, D and Hand, V. "Nanoparticles in the Environment: Can we learn from the Mineral World?" School of Earth, Atmospheric and Environmental Sciences, and Williamson Research Centre for Molecular Environmental Science, University of Manchester, U.K. www.nanomat.de/pdf/nanovision-vaughan.pdf (Accessed on January 14, 2011)

2. http://www.understandingnano.com/ index.htm (Accessed on January 14, 2011)

Chapter 1
An Introduction to the Scope of Nanoscience and Nanotechnology

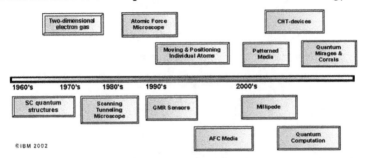

Figure 1. Timeline for the development of nanoscience and nanotechnology
(Resource: IBM nanotechnology research webpage, 2002).

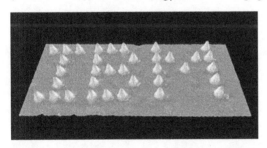

Figure 2. IBM's logo, designed with 35 Xenon atoms
(Resource: IBM nanotechnology research webpage, 2009).

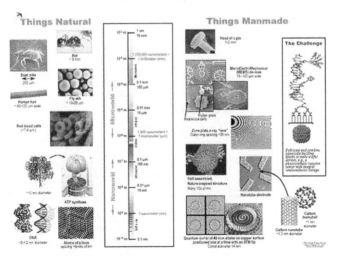

Figure 3. Natural and artificial objects at the nanometer scale
(Resource: Office of Science, US department of Energy, May 2006).

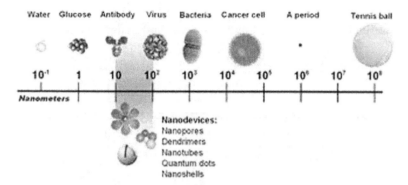

Figure 4. The nanometer scale defining different living or non-living entities (Resource: National Cancer Institute).

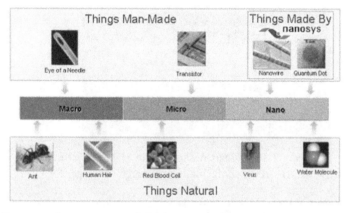

Figure 5. The nanometer scale objects occurring naturally or artificially in nature (Resource: Nanosys, Inc.)

Figure 6. Polychrome lusterware cup from ancient (9th Century AD) Iraq (Resource: Photograph from the collections in the British Museum).

Figure 7. The Lycurgus Cup originally in green color turns red in reflected light (Resource: Photograph from the collections in the British Museum).

Figure 8. Sword of the Indian King Tipu Sultan, made from Damascus Steel.

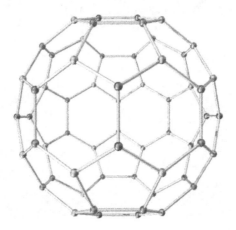

Figure 9. Buckyball or buckminsterfullerene (a carbon-60 soccer ball) (Resource: POV-Ray 3D animation).

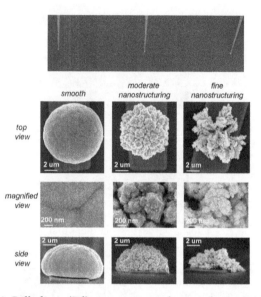

Figure 10. Palladium (Pd) nanostructured microelectrodes (NME) can act an advanced nanobiosensor and have been used to determine nanotexturing on the same chip. (Ref. Soleymani, L., Fang, Z., Sargent, E.H., Kelley, S.O. Programming the detection limits of biosensors through controlled nanostructuring. Nat. Nanotechnol. 4, 844 - 848 (2009).

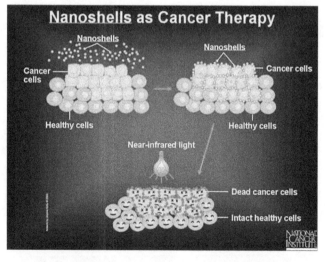

Figure 11. Light mediated cancer cell detection using engineered nanoshells (Resource: National Cancer Institute).

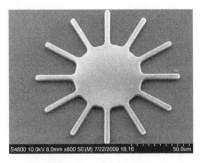

Figure 12. Microelectromechanical systems (MEMS) contains tiny moving parts, which can filter or vibrate with an electrostatic force upon receiving the incoming signal (Resource: Chivukula, V.B. Purdue University, USA)

Figure 13. An imaginary structural artwork of MEMS
(Resource: www.memx.com/technology.htm)

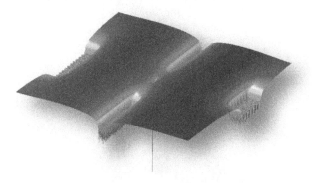

Figure 14. A single atom nanotransistor for quantum computers
(Resource: Science Daily, 2012).

1 cm

Figure 15. Hybrid-Insect Miceroelectromechanical System (HI-MEMS) was proposed as a at the U.S. Defense Advanced Research Projects Agency (DARPA) to create robots as complex as insect forms that required millions of years of evolution to achieve, scientists now essentially want to hijack bugs with electronics to create "hybrid insect vehicles" (Resource: Scientific American blog 2012).

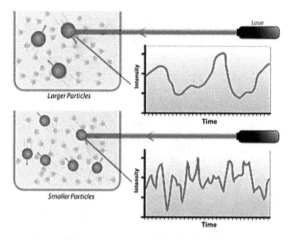

Figure 16. Size dependent movement of the particles within the same time frame.

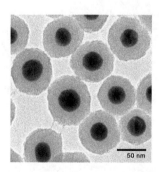

Figure 17. Transmission Electron Micrograph (TEM) of silica capped gold nanoparticles (Resource: http://nanocomposix.com/pages/gold-colloid)

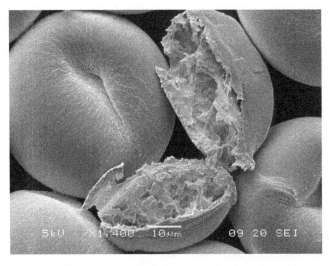

Figure 18. SEM of Polylactic Acid Shell (Resource: American Scientific, 2009).

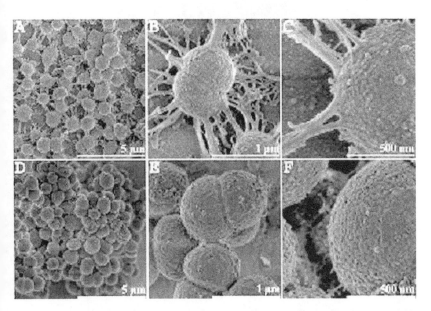

Figure 19. FESEM images of sessile (A, B, and C) and planktonic (D, E, and F) cells of D. geothermalis E50051 (Reference: Saarimaa, C., Peltola, M., Raulio, M., Neu, T.R., Salkinoja-Salonen, M.S., Neubauer, P. Characterization of Adhesion Threads of Deinococcus geothermalis as Type IV Pili. J. Bacteriol. 188 (18), 7016-7021, 2006).

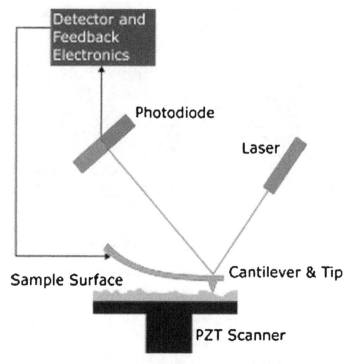

Figure 20. Schema for AFM (Resource: Casa, S.D. 2012)

Chapter 2

NATURAL NANOPARTICLES

R. L. Brahmachary

Nanoparticles, inorganic and organic, are widely distributed in nature. Nanoparticles may be formed when volcanic eruptions spew out materials from below the surface of the earth or even when tidal waves or swift currents of rivers erode a coast or bank. The notorious eruption of Eyjafjallajokull injected in the atmosphere nanoparticles of the size range 100-400 nm (Harrison 2010 et al). Hochella and his collaborators (2008) opine that the role of Fe^{+2} hydrate nanoparticles in transporting heavy metals down the Clark Fork river, Montana, USA, is quite significant, and that the nanoparticles are of the size range 5-10 nm (Rebecca French, personal communication). Banfield and Zhany (2001) described nanometer scale porosity that separates oriented nanocrystals in certain geological samples.

It was known decades ago that certain bacteria orientate themselves along the magnetic north-south thanks to a magnetic needle they contain, but only as late as 1987 was it established that the needle is composed of a "collection" of nanosized iron particles (Bellini, S,. 2009). Likewise, the beautiful silica-based structures in diatoms were also found to be formed of nanosize silica units. (Bradbury J 2004): Abracado et al (2005) reported ~11-12 nm-sized magnetic particles in the head, thorax and abdomen of a Brazilian ant, *Solenopsis substituta*. These particles probably help the ant to orient itself in the terrestrial gravitational field. They also list a number of ant species which orientate themselves in a magnetic field. Acosta Avalos isolated magnetic nanoparticles from the ant *Pachycondyla marginata*. Zhang et al (2006, 2007) detected titanium and calcium in certain fish, but this is a case of bioaccumulation and has been clearly established in the pigmentless medaka (see-through medaka). Eggs of this transparent medaka on exposure to ~40 nm fluorescent particles accumulated these external objects. These nanoparticles made their way into the gall bladder and yolk region of the post-hatch larvae. These facts suggest that living organisms are likely to take up many environmental nano-objects, and these should be treated as contaminants rather than natural nanoparticles. However, if nanoparticles are functional in an organism, as are magnetic, in ant (or bacteria), we may consider these as "natural" nanoparticles. Bees, birds, dolphins etc. have also been reported to possess magnetic particles, but some of these may not be of nano size.

Many molecules playing important roles in living cells are nanosized; in other words, these are natural nanoparticles. We know that proteins such as antibodies are ~10 nm, lysozyme is 4 nm, and BSA is 5-10 nm (Fig. 3). Furthermore, a living system contains nucleic acids, lipids, carbohydrates and other compounds having specific functions, and their dimensions are to be considered. Table 1 furnishes certain relevant data.

Table 1. Approximate sizes (in nm) of certain molecules which play important roles in the living system

DNA diameter	2
Glucose	1
Amino Acids	1
Proteins	1-20
Specific Examples	
Antibodies	10
Bovine Serum Albumin	5-10

Insulin	5
Hemoglobin	5.5
Earthworm Hemoglobin	30

With this introduction, we move to our principal task i.e. developing an approach that sheds light on phylogenetic evolution and adaptation to environment of natural nanoparticles. We are collecting experimental data relevant for the above two topics. The data, far from exhaustive and gathered with the help of a technique Dynamic Light Scattering (DLS)-that has its limitations-nevertheless indicates two trends which are of interest to the evolutionary biologist.

After numerous attempts, we have established that despite shortcomings DLS yields results which are reliable so far as our approach is concerned (For detailed descriptions of DLS refer to Chapter 1 of this book). Unless the particles are exactly spherical, the different dimensions along different axes will necessarily yield different values for the same particle. Many natural nanoparticles are indeed not perfect spheres. Scanning Electron Microscopy and Transmission Electron Microscopy can then resolve doubts as to the structure. A large number of size-classes of nanoparticles i.e. a high degree of polydispersity would prevent obtaining meaningful results. Furthermore, suspended large particles, pigmented and/or fluorescent material, would totally mask the results. Repeated centrifugation and filtration are sometimes helpful. 10-20 repetitions of the same sample reading were often undertaken in order to determine meaningful results. Furthermore, viscous or dense solutions have to be diluted many times. The following results obtained with a pure substance like the protein Bovine Serum Albumin (BSA) illustrate this point. Figure 1 shows the peak of distribution at 6 nm, and this value is in agreement with table 1.

Results based on 15 readings of the same samples, i.e. the data obtained from 14 repetitions of DLS operation of the same sample (5 mg was dissolved in 4 ml PBS buffer), shows that the sample belongs to a size class of 10.2-11.3 nm, and this is in agreement with the known size class of BSA.

After 20 days in solution, 15 readings of the sample revealed an aggregation or loosening of the structure, resulting in a very large increase in the dimensions of the particles. They now ranged in the size class ~160-187 nm. This trend was verified with another old solution. In fact, even after only 4 days in solution, the particles showed a peak at ~220 nm; the polydisperse nature was also soon revealed.

Later, we studied silkworm hemolymph nanoparticles in solution and detected a time-dependent increase in size; apparently, the reverse process i.e. decrease in size (through fragmentation) does not occur.

Natural Nanoparticles in Living Systems

(a) *Animal system*: Hemolymph of silkworm, sera of certain vertebrates and human saliva were investigated. The material was diluted in PBS buffer and filtered repeatedly when so required for yielding data meeting the quality criteria of DLS. For saliva, both water and PBS were used. PBS data show a slightly higher size of nanoparticles.

(b) *Plants*: Different parts of plants such as leaves and stems were washed and then crushed with mortar and pestle. The extract was diluted in Millipore water and then centrifuged repeatedly, generally at 14,000 rpm for 10 minutes till it was free of significant amounts of sediments. In order to remove pigments such as chlorophyll, repeated dilutions were often necessary.

(c) *Mushroom* was also subjected to a similar procedure.

Table 2 sums up the data obtained on the animal systems. In hemolymph of silkworm, a large number of samples show a distinct pattern; 60-80% of the nanoparticles at 10-15 nm size class claim 60-80% intensity; the rest belong to the 200-300 nm size class. Since the silkworm was constantly feeding on mulberry leaves, nanoparticles in this food source were also investigated. The major peak of leaf extract (~91% intensity) is claimed by a 184 nm peak while the rest (20. 3nm) corresponds to 5.3% intensity, and so the size class (10-15 nm) which is most prominent in hemolymph (60-80%) is not directly due to the food source of the silkworm.

The nanoparticles in chicken serum (6 samples) reveal a picture that is almost the reverse of that of hemolymph, namely, the 11 nm size class claims only 5-7% intensity, while the major peak at 105-118 nm answers for more than 95% intensity. As evident from the table, the pattern is similar in human sera (12 nm: 9%; 115 nm: 88%). In a race horse just after retirement, the major peak (81%) lies at the 52 nm size class and the second peak (19%) corresponds to the 10 nm size class. Unfortunately, we had only a single sample, but we have briefly discussed later the possible significance of this and pointed out why a further investigation with more samples is warranted.

Table 2 Nanoparticle distribution of some animal species

Sample Name	No. of Samples	Particle Size Class (nm)	Intensity (%)
silkworm haemolymph	numerous	10-15	60-80
		200-300	25-30
chicken serum	6	11	5-7
		105-118	90-100
horse serum	1	10	19
		52	81
human serum	5	14	7
		118	90
human saliva	2	118	94

Table. 3 Nanoparticle distribution of some plant species and one mushroom species

Sample name	Particle Size Class (nm)	Intensity (%)
Sporobolus (grass)	30	0.9
Sporobolus (grass)	180-205	97-98
Sporobolus (grass)	200	100
Eleusina aegyptica (grass)	164-194	98-100
Colocasia (monocot)	217-251	90-97
	micron size	3-10
Oleander latex	183-214	99
Sonchus (dicot)	16	3-5
Sonchus (dicot)	97-111	90-98
Mung bean (*Vigna radiata*) Seedlings	18	2
Mung bean (*Vigna radiata*) Seedlings	135-144	95
Blue green algae (pond water)	43-51	2-4
Blue green algae (pond water)	197-253	95-98
Agaricus augustus	140-180	100

Human saliva (Fig. 1) shows the major peak at 105-118 nm; the smaller peak at ~16nm is ~6%.

We thus see that in the circulatory system of the animal world, the 10-15 nm size class appeared early in the course of evolution, namely in invertebrate haemolymph, but this small and therefore probably more reactive group of nanoparticles then largely decreased. In the vertebrate sera, the small size class has spectacularly decreased at the expense of the 100+ nm size class. It is possible that insect larval haemolymph which lacks the specific immunological reaction of vertebrates stores a large quantity of general purpose antibodies (Rosengaus 1999). In the race horse, the major peak is relatively high — the 52 nm size class claims the lion's share.

Apart from phylogenetic evolution, the metabolism of animals adapted to their specific lifestyle (such as the difference between a homing pigeon and a hen) may be reflected in the nanoparticles in their sera. It would therefore be worthwhile to pursue this quest by comparing domestic poultry birds like fowl and duck with those of wild duck, teal or goose, which undertake long regular flights in the course of their normal lifestyle. However the logistics of capturing such birds and extracting their blood is enormously difficult, and perhaps in the near future homing pigeons and horses can be investigated in greater detail.

Table 3 shows the picture in the plant world. Mushroom, *Agaricus augustus*, has been included in it. The general impression is that in plant extracts (leaf extract, sap from stem etc.), small size particles are meager in quantity, and with the exception of *Sonchus*, the major peak corresponds to a size class significantly larger than that of vertebrate sera. In plants, blue green algae seem to be on the higher side (197-253 nm: 95-98%), while in *Colacasia* sap, the 217-251 nm size range answers for the 90-97% size class.

Table 4 shows the intriguing results obtained with blue-green algae of a local pond, compared with such algae when adapted to 70°C water from a thermal spring (Taptapani,Orissa,India).

Table. 4 Nanoparticles (in nm) in pond algae and in those adapted to 70⁰C water of a thermal spring

Sample	Particle size	Intensity%
Blue green algae(from pond)	197-253	95-98
Blue green algae(from pond)	43-51	2-4
Bluegreen algae(from thermal spring)	614-714	98-100
Bluegreen algae(from thermal spring)	33	1

Although the exact species of the algae were not known – and often more than one species combine in a colony – the spectacularly higher size class of the thermal-spring-adapted algae makes possible an explanation based on the altered metabolism of the organism that has evolved to cope with this unusual environment.

This is perhaps corroborated by the results of an *in vivo* heat shock experiment (Tables 6 and 7) we performed with the grass *Sporobolus*. In the control, the leaf of a grass clump was cut off and the clump (in a pot) was subjected to 5-hour shock at 42^0C. The major peak in the control leaf was at 217 nm, while after the 5 hour heat shock, the major peak was at 359 nm. In a repeat experiment, the values were 203 nm and 289 nm, respectively. Similarly, heat shock for 2 hours at 50^0C increased the size. We are immediately reminded of heat shock proteins, but at present we do not know the chemistry of these nanoparticles. Again, a 20 hr shock leads to a profile too polydisperse to record. In nature, the thermal-spring-adapted algae maintain their profile.

We thus obtain two results of evolutionary importance:

1. In haemolymph, the major peak (60-80%) of nanoparticles belongs to the small 10-15 nm. In vertebrate sera, there is a sudden reversal of the picture: the major size class is 100+ nm, though in the race horse it is 52 nm(but not as small as 10-15 nm)

2. The heat-adapted blue-green algae show much larger nanoparticles than those of the pond algae (614-714 nm as opposed to 197-253 nm).

**Table. 5 Increase in Size of Nanoparticles in *Sporobolus*
due to Heat Shock (5 hours at 42°C)**

		Size in nm	Intensity%
Experiment 1	Control leaf at 30^0C	217	95%
	Heat shock for 5 hours at 42^0C	319	95%
Experiment 2	Control as above	203	95%
	Heat shock as above	289	95%

Table. 6 Increase in Size of Nanoparticles in *Sporobolus*
due to Heat Shock (2 hours at 50°C)

Experiment 1		Size in nm	Intensity%
	Control leaf ~30°C	179	100%
	Heat shock at 50°C	268	95%
Experiment 2	Control leaf ~30°C	206	94%
	Heat shock 50°C	269	97%

Ontogenetic Evolution

Ontogenetic evolutionary significance is highlighted in the silkworm example, where the larval-stage small size class predominates only in the larval stage but not in eggs or adults (moth).

Silkworm eggs were also extracted and subjected to DLS. The major peak turned out to be in the 148-194 nm range (94-98%). An example is: 148.3 nm (95.7%), 36.2 nm (1.7%), 5484 nm (2.5%). Likewise, when haemolymph from an adult silk moth was extracted, the major peak turned out to be 140-186 nm, 152.1 nm (90%), 15.3 nm (3.2%), 4851 nm (7%). Thus only in the larval stage the small size class predominates.

SEM Imaging

The first attempt at imaging haemolymph nanoparticles was not successful; the high voltage created artifacts. At 5kV, the image of nanoparticles in the foxtail grass leaf extract is quite clear.   It is evident (Fig. 4) that particles, mostly rectangular, have different size dimensions along the major axes. Furthermore, the particles belong to significantly different size classes. With another image at a lower magnification we confirm the presence of particles ranging from 221 to 392 nm, as measured along the longer axis. DLS yields a peak at ~200 nm but, the peak being broad, the range extends from about 100 nm to more than 300 nm. Considering all its limitations, DLS has indeed proved its worth.

Actinomycin Treatment

100-200 microgram Actinomycin D solution was smeared on foxtail leaves. 3 hours following this treatment, the nanoparticles in the leaves were tested again with DLS, and no alteration was detected. 3 days after smearing the leaves, extraction, followed by DLS examination, was again carried out. A significant difference was noted. A peak was obtained at 612 nm, compared to the control (untreated leaf), which indicated a peak at 225 nm.It is difficult to explain the results. Actinomycin may inhibit protein synthesis via RNA synthesis, and some of these proteins and or enzymes required in maintaining homeostasis might have been missing. The results are rather similar to the aggregation of nanoparticles forming higher sized particles with time, when left in a solution in the test tube.

Addendum

(i) Cold shock (4.5 hrs at 4⁰C) increased the *Sporobolus* nanoparticles size from 182 nm to 203 nm i.e. only very slightly, unlike the heat shock.

(ii) Aqueous extract of *Camponotus sp.* ant , locally obtained, revealed two or three size classes. Representative example is ~290 nm (~88%), ~56 nm (8.5%).

iii) *Aspergillus niger* and Sulphur: Sulphur has been found to have an effect on *A. niger* spores. Post-treatment studies reveal that the size is greatly reduced (Roy Choudhury et al). This is not unexpected, because *A.niger* is inhibited or killed by sulphur, particularly nanosulphur, and so the fungal system no longer manifests the normal metabolic pattern.

Acknowledgements

All the experiments described here have been performed at the Nanobiotechnology lab headed by Prof. A. Goswami, AERU, ISI, Kolkata, with his active interest. We thank Mojammel Haque Mandal and Dr Manabendra Mukhejee of SINP, who took the trouble to measure a large number of samples before we acquired our DLS, and for stimulating discussions we had at their lab. Apart from the regular research scholars at this lab – Ayesha Rahman, Dipankar Seth, Samrat Roy Choudhury – short term project workers Joyita Bhadra, Rochishnu Dutta, Arurnima Gupta   and Indrani Roy also gave their time and their effort toward completing this project. Samples of heat adapted algae were a kind gift from the lab of Dr. Anjana Dewanji, A.E.R.U, ISI.

References

1. Abracado,L.G. (2005)   D.M.S. Esquivel, O.C. Alves, and E. Wajnberg: Magnetic Material in head, thorax and abdomen of Solenopsis substituta ants: A ferromagnetic resonance study. Journal of Magnetic Resonance 175: 309-316.

2. Acosta-Avalos D. et al (1999) Isolation of Magnetic Nanoparticles from Pachycondyla marginata ants. The Journal of Experimental Biology 202:2687-2692.

3. Banfield J,.F,Zhang (2001) H, Nanoparticles in the Environment,Reviews in Mineralogy and Geochemistry Volume 44: 1-51

4. Bellini,S. (2009) On a unique behavior of freshwater bacteria. Chinese Journal of Oceanology and Limnology Vol. 27 (1): 3-5,

5. Bradbury J (2004) Nature's Nanotechnologists: Unveiling the Secrets of Diatoms. PLoS Biol 2(10): e306. doi:10.1371/journal.pbio.0020306)

6. Harrison, R G ,et al. (2010) Self-charging of the Eyjafjallaj¨okull volcanic ash plume. Environ. Res. Lett. 5: 024004-8

7. Hochella Jr. Michael F. et al (2008) Nanominerals, Mineral Nanoparticles, and Earth Systems Science Vol. 319 (5870): 1631-1635

8. Rosengaus RB, Traniello JFA, Chen T, Brown JJ and Karp RD (1999). Immunity in a social insect. Naturwissenschaften, 86:588–591

9. Roy Choudhury S, Ghosh M, Mandal A, Chakravorty D, Pal M, Pradhan S, Goswami A. (2011) Surface-modified sulfur nanoparticles: an effective antifungal agent against Aspergillus niger and Fusarium oxysporum. Appl.Microbiol Biotechnol. Apr; 90(2):pl 733-43.

10. Zhang XZ, Sun HW, Zhang ZY. (2006) Bioaccumulation of titanium dioxide nanoparticles in carp Huan Jing Ke Xue Volume 27 (8): 1631-5 in. TERI project: Capability, Governance, and Nanotechnology Developments - a focus on India New Delhi: The Energy and Resources Institute.[Project Report No. 2006ST21: D3] The Energy and Resources Institute (TERI). 2008 A report on risks from a developing country perspective.

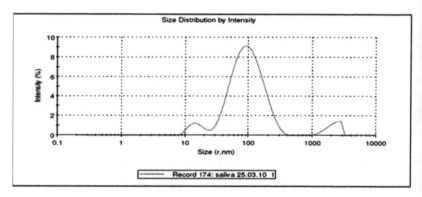

Figure 1. Dynamic Light Scattering (DLS) data representing the hydrodynamic radius of Human Saliva (Resource: Unpublished data from Prof. Ratanlal Brahmachary).

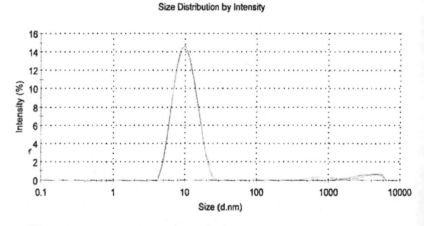

Figure 2. Dynamic Light Scattering (DLS) data representing the hydrodynamic radius of BSA protein molecules
(Resource: Unpublished data from Prof. Ratanlal Brahmachary)

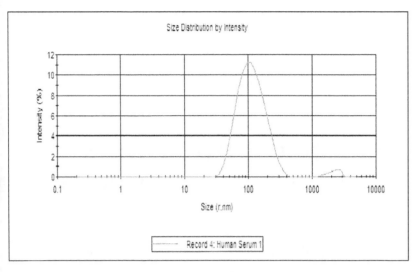

Figure 3. Dynamic Light Scattering (DLS) data representing the
hydrodynamic radius of Human Serum
(Resource: Unpublished data from Prof. Ratanlal Brahmachary).

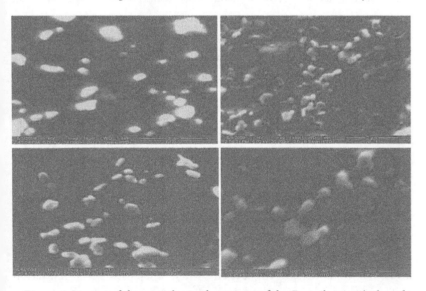

Figure 4. Images of the particles in the extract of the Foxtail grass. (a, b and
c) the particles at scale of 2 μm. (d) the particles at the scale of 1μm (Resource: Unpublished data from Prof. Ratanlal Brahmachary).

Chapter 3

BIOLOGICAL IMPLICATIONS OF METALLIC NANOPARTICLES

Shilpi Banerjee, Dhriti Ranjan Saha and Dipankar Chakravorty

Nanoscience as we know it today was ushered in by the study of metal nanoparticles. A pioneering contribution was made by Michael Faraday, who explained the various colors exhibited by a gold colloidal solution as arising due to the presence of the phase "in a metallic divided state."[1] Optical absorption of metal nanoparticles dispersed in a dielectric medium arises due to Surface Plasmon Resonance.[2]

Various methods have been used to prepare nanosized metal particles. In inert gas condensation technique, the metal is evaporated in an inert gas and the atoms are condensed on the surface of a cold finger. The nanoparticles formed are then scraped off, collected in a die and pressed under high vacuum to form a nanocrystalline metallic pellet.[3, 4] Chemical methods have been used extensively to form metal nanoparticles. The simplest route is to reduce a metallic salt with a suitable reducing agent.[5] A sol consisting of a silicate glass precursor and the metallic salt concerned, when subjected to a reduction treatment, leads to the formation of metal nanoparticles within the glass matrix.[6] Alkali silicate glasses, after being ion-exchanged with silver or copper ions and then subjected to a reduction treatment in hydrogen, produced nanoparticles of the metal phase concerned.[7, 8]

Reduction of an ion-exchanged glass-ceramic also led to the growth of metal nanoparticles.[9] Electrodeposition technique has been shown to give rise to a fractal growth of metallic structure within a glassy electrolyte. The former consists of metallic nanoparticles.[10,11]

Extensive studies on electrical, magnetic and optical properties of metallic nanoparticles have been reported in the literature.[12] Of late, these particles have assumed importance in the characterization of biological systems. In this chapter we briefly review the progress made in the applications of metallic nanoparticles to biological systems. The synthesis techniques used to make metal nanoparticles and the exploitation of biological systems to prepare them are also discussed.

1. Synthesis

1.1 Synthesis of Silver Nanoparticles using Electrochemical Process:

Silver nanoparticles are synthesized by electrochemical technique.[13] The chemicals used in the experiment were: silver nitrate ($AgNO_3$), ethanol (99.8%) and 99.99% metallic silver. Silver polarization was performed in a saturated solution of $AgNO_3$ in ethanol de-aerated by argon. Using PGZ301 Voltalab potentiostat by means of potentiodynamic (Cyclic voltammetry) and potentiostatic technique (chronoamperometry), anodic silver properties were investigated at room temperature. Here silver rod was used as a working electrode, a platinum plate as a counter-electrode and Ag/AgCl as a reference electrode.

1.2 Gold Nanorods prepared by Electrochemical Method:

Gold nanorod was synthesized within a simple two-electrode type cell by electrochemical oxidation/reduction.[14] A gold metal plate was used as the anode and a platinum plate used as the cathode. The electrodes were immersed in an electrolytic solution of hexadecyltrimethylammoniumbromide ($C_{16}TAB$, 99%; Sigma), and a rod-inducing cosurfactant. $C_{16}TAB$ was used as supporting electrolyte and also to stabilize the nanoparticles to prevent their further growth. In this process the bulk gold metal was converted from the anode, and formed gold nanoparticles most probably at the interfacial region of the cathodic surface and within the electrolytic solution in a controlled-current electrolysis for a typical current of 3 mA, a typical electrolysis time of 30 minutes under ultrasonication, and at a typical temperature of 311 K.

1.3 Synthesis of Silver Nanoparticles with Different Shapes using Capping Agent:

The range of applications for metal nanoparticles can be increased by controlling their shapes. Silver seed was synthesized by using silver trifluoroacetate (CF_3COOAg) as the source material. Silver nanoparticles with different shapes were synthesized by a seed growth method.[15] In order to obtain different shapes of silver nanoparticles, all conditions like seed, precursor, temperature, reductant and concentrations of the reagents were retained, and only the capping agent was changed. In this way, using spherical Ag seed, different shapes of nanosilver like octahedrons, cubes and bars etc. could be prepared. Fig. 1 shows the transmission electron micrographs for different shapes of silver nanocrystals grown by Zeng et al.[15]

1.4 Cylindrical Gold Nanorods Synthesis using Wet Chemical Method:

Gold nanorods with different aspect ratios like of 4.6±1.2, 13±2, 18±2.5 (with 16±3nm short axis) were prepared using a seeding growth method.[16] 3.5 nm-diameter gold particles were used as seed. The seed was prepared by the reduction of $HAuCl_4$ with borohydride, with sodium citrate as the capping agent. By varying the ratio of seed to metal salt, the aspect ratio of the nanorods could be varied. The long rods were separated from the spherical particles and the surfactants by centrifugation.

1.5 Synthesis of Gold Nanorod using Seed-mediated Growth Method:

Gold nanorods were prepared using seed-mediated growth method.[17] Here hexadecyltrimethylammonium bromide (CTAB) was used as the capping agent. In a single-component surfactant system using silver-containing growth solution, gold nanorods of desired length with aspect ratios varying from 1.5 to 4.5 were synthesized. Using binary surfactant system with benzyldimethylhexadecylammoniumchloride (BDAC) and CTAB respectively, longer nanorods with length 4.6 to 10 nm could be prepared.

1.6 Synthesis of Gold and Silver Nanoparticles using Bacteria Bacillus Subtilis:

Using the bacteria *Bacillus subtilis,* gold and silver nanoparticles have been synthesized.[18] Aqueous solutions of $HAuCl_4$ and $AgNO_3$ were taken as sources of gold and silver respectively. Broth solution containing *B. subtilis,* after 72 hours of culture, was mixed with $HAuCl_4$ or $AgNO_3$ to obtain gold and silver nanoparticles respecrively. Gold nanoparticles were synthesized both intra- and extra-cellularly. Diameters of gold nanoparticles were 7.6 1.8 and 7.3 2.3 nm respectively, in intra- and extra-cellular method. At the same time, silver nanoparticles with diameter 6.1 1.6 nm were prepared extra-cellularly. The gold nanoparticles were also formed after one day of addition of chloroaurate ions, while the silver nanoparticles were formed after seven days.

1.7 Using Apiin as a Reducing Agent in the Synthesis of Gold and Silver Nanoparticles:

The biological synthesis of anisotropic gold and quasi-spherical silver nanoparticles were reported.[19] In this case, chloroauric acid trihydrate ($HAuCl_4 \cdot 3H_2O$) and silver nitrate ($AgNO_3$) were taken as the sources of metal. Here apiin, a chemical compound isolated from parsley and celery, was used as both the reducing and stabilizing agent. The size and shape of the nanoparticles was controlled by controlling the ratio of metal salts to apiin compound. From electron microscopy, it was found that the average sizes of the gold and silver nanoparticles were 21 and 39 nm respectively. Due to surface binding of apiin compound with the reduced material, the synthesized nanoparticles remained stable in water for more than three months.

1.8 Biosynthesis of Silver and Gold Nanoparticles using Phyllanthin:

Gold and silver nanoparticles were synthesized using aqueous chloroauric acid ($HAuCl_4$) and silver nitrate ($AgNO_3$) solutions as sources of gold and silver respectively.[20] Reducing agent phyllanthin was collected from the plant *Phyllanthusamarus*. Aqueous source solution was reduced by the extract of phyllanthin at room temperature, to give anisotropic gold and spherical or quasi-spherical silver nanoparticles. It was found that, to reduce $AgNO_3$, a larger amount of phyllanthin extract was required than that for $HAuCl_4$. The concentration of phyllanthin extract controlled the size and shape of nanoparticles.

1.9 Fungus-assisted Synthesis of Silver Nanoparticles:

Silver nanoparticles were synthesized using the fungus *Verticillium*, which was collected from the plant *Taxus* and then maintained on potato dextrose agar slants at 298 K.[21] Aqueous $AgNO_3$ was reduced to silver nanoparticles when it was exposed to the intracellular fungal biomass. Electron microscopy revealed that silver particles were formed below the cell wall surface due to reduction of the silver ions by the enzymes in the cell wall membrane. The size of the silver nanoparticles was around 25, ± 12 nm. The cells continued to multiply after biosynthesis of the silver nanoparticles because the metal ions were not toxic to the fungal cells.

1.10 Synthesis of Silver Nanoparticle with the help of Plant Leaf Extracts:

Silver nanoparticles were prepared by reducing aqueous solution of $AgNO_3$, using plant leaf extracts from different species as reducing agents.[22] Plant leaves from Pine (*Pinus desiflora*), Persimmon (*Diopyros kaki*), Ginkgo (*Gingko biloba*), Magnolia (*Magnolia kobus*) and Platanus (*Platanus orientalis*) were collected and air-dried at room temperature for two days. Using the UV-visible spectroscopy, the quantitative formation of silver nanoparticles was monitored. Experiments revealed that, in terms of synthesis rate and conversion into silver nanoparticles, Magnolia (*Magnolia kobus*) leaf broth proved to be the best reducing agent. Using Magnolia leaf broth, 11 minutes was required for the conversion of 90% at the reaction temperature of 368 K.

1.11 Biosynthesis of Metal Nanoparticles using Cloves (Syzygium aromaticum) as Reducing Agent:

In this exercise, natural precursor clove was used as the reducing agent to prepare metallic nanoparticles (Au and Ag) by a simple, cost-effective method.[24] Clove extract was prepared by dipping clove in water for 24 hours and then filtering it. Aqueous solutions of $AuCl_4$ and AgNO3 were used as metal sources respectively. In this exercise, the reduction time required was very small. The shape and size of metal nanoparticles were controlled by varying the ratio of metal salts and reducing agent.

1.12 Green Synthesis of Metal Nanoparticles:

In the green chemistry preparation of nanoparticles, the solvent medium used for the synthesis, an environmentally benign reducing agent, and nontoxic material for the stabilization of the nanoparticles should be considered. In this case $AgNO_3$ solution was used as the source of silver. The environmentally benign solvent, water, and the reducing sugar β-D-glucose was taken as the reducing agent as it was inexpensive and nontoxic.[23] Starch was selected as the capping material to passivate the nanoparticle surface for producing stable nanoparticles.

1.13 Synthesis of Water-soluble Silver Nanoparticles:

In this exercise, silver metal nanoparticles (Ag NPs) were synthesized using $AgNO_3$ and glutathione (GSH).[25] The synthesized silver nanoparticles were size-tunable and water-soluble. These silver nanoparticles could be bound covalently with bovine serum albumin, which is a biocompatible protein, and also with other functional molecules in mild conditions.

1.14 Synthesis of Gold, Silver Nanoparticles and Gold Core Silver Shell using Neem Leaf Broth:

In this method, metallic silver and gold nanoparticles and bimetallic Au/Ag nanoparticles were synthesized in extra-cellular process using neem (*Azadirachta indica*) leaf broth as the reducing agent.[26] Aqueous solutions of silver nitrate and chloroauric acid were taken as source of metal. Stable silver and gold nanoparticles were formed rapidly at high concentrations. The silver and gold nanoparticles which formed were polydisperse. For the synthesis of bimetallic Au core–Ag shell, gold and silver sources were taken in the solution and then exposed to neem leaf broth. Transmission electron microscopy revealed that the silver nanoparticles were adsorbed onto the gold nanoparticles during synthesis of the bimetallic Au core–Ag shell particles. This method was by far the fastest amongst all methods involving plant extracts as reducing agents.

2. Application

Nanotechnology offers newer and superior solutions for challenges like diagnosing, imaging, drug delivery and also for engineered materials as replacements for damaged tissues. The unique size-dependent properties of nanomaterials make them applicable in many areas of biomedical research and medicine like:

- Biological labels
- Biosensor
- Drug delivery
- Detection of proteins
- Tumour destruction via heating (hyperthermia)
- Biomedicine

Since nanoparticles and proteins are in the same size domain, nanomaterials are suitable for bio-labeling. The biocompatibility of nanoparticles can be greatly enhanced by coating them with antibodies, biopolymers and other suitable small molecules.

2.1 Biological Tagging of Antibody-conjugated Gold Nanoparticles:

Here is a brief description of the exercise in which biological tagging of antibody-conjugated gold nanoparticles of resected human pancreatic cancer tissue and its subsequent optical detection was carried out.[27] The gold nanoparticles were prepared from 15 nm spherical gold cores and then stabilized by heterobifunctional polyethylene glycol (PEG). Next, the stabilized gold nanoparticles were covalently coupled to F19 monoclonal antibodies. Here, heterobifunctional PEG ligands made the gold nanoparticles stable as it contained a dithiol group. Its terminal carboxy group was used for the coupling of antibodies outside of the PEG shell. It was found that the gold nanoparticles and the antibody bioconjugates formed a stable dispersion and did not agglomerate for a long time. The synthesized nanoparticle bioconjugates were used as biological probes to label tumor stroma inside 5 μm-thick sections of resected human pancreatic adenocarcinoma. The tissue samples were observed using dark-field microscopy near the nanoparticles resonance scattering maximum which approximately 560 nm. From the images, the pronounced changes in the cancerous tissue features could be observed. This study suggested that this labeling method, using gold nanoparticles and the antibody bioconjugates, was effective as a rapid and accurate method for the identification of cancer tissue.

2.2 Use of Gold Nanoparticles in PIC Imaging:

In this exercise, membrane proteins in cells were visualized or imaged using 10 nm gold particles and an all-optical method.[28] Here, both the transverse and axial resolution of the Photothermal Interference Contrast (PIC) method was measured and then it was compared to the measurements of a simple model. The PIC method is more effective than fluorescent methods for single-molecule detection for the following reason: saturation of the signals, autofluorescence or scattering caused by the environment or the cells themselves, and blinking and photobleaching of the labels were all absent in the PIC method. In performing three-dimensional (3D) localization of a single fluorescent molecule, the blinking and photobleaching of the labels should be absent since these made the records of the molecule hardly feasible. In the case of PIC, signals localized the 3D single particles and gave a large number of recordings. Also, in this case, the signal levels were arbitrarily high. Therefore, in this process, single nanoparticles could be localized in the scattering environment with large pointing accuracy formed by a cell. This exercise established that, in principle, it was possible to multiplex fluorescence and also in PIC images. With the PIC method, the image spot was located more accurately than what was possible by specific cellular organelles, fluorescently labeled. Nanoparticles with a shifted plasmon resonance arising due to different shapes or composition could be taken for multiplecolor PIC imaging. In the case of live biological samples, the signal-to-noise ratio (SNR) should be limited by the temperature rise in the sample. This temperature rise decreased reciprocally with the distance from the particle center. The extremely low number (one or two) of 10 nm gold colloids present on IgG antibodies could be estimated using the high sensitivity of the method. In the case of low-expression protein detection, the technique was very useful. The method gave an unbiased picture of the expression pattern of the cell without any amplification. For the stoichiometry of gold nanoparticles which were commonly used in protein and DNA chips, this method was very suitable.

2.3 Application of Silver Nanoparticles in Neuroblastoma Cell as Biological Labels:

The unique plasmon-resonant optical scattering properties silver nanoparticles (SNPs) allow SNPs to be used as signal enhancers, optical sensors, biomarkers etc. Here is a description of the potential of silver nanoparticles in neuroblastoma cells as biological labels.[29] Ag nanoparticles of similar sizes and with different surface chemistry (hydrocarbon and polysaccharide) were studied from the point of view of biological labeling. A strong optical labeling of the cells was observed for the excitation of plasmon resonance of both types of SNPs in a high-illumination light microscopy system after 24 hours of incubation. Using scanning electron microscopy (SEM), the surface binding of both types of SNPs to the plasma membrane of the cells was observed. With the help of transmission electron microscopy (TEM) the internalization and localization of the Ag nanoparticles inside the intracellular vacuoles of the thin cell sections were observed. With Ag nanoparticles at concentrations of 25 μg ml^{-1} or greater, a nerve growth factor was found after incubation.

2.4 Gold-silica Core-shell Nanorods: A Promising Material for Molecular Photoacoustic Imaging:

In this exercise, gold-silica core-shell nanorods were synthesized.[30] The shell thickness of silica could be controlled. Using UV-Vis spectroscopy, electron microscopy, and FDTD numerical simulation, their stability in aqueous solution to nanosecond high-energy laser pulses at the surface plasmon resonance was studied. Silica-coated gold nanorods were optically more effective than the bare gold nanorods. Under high-energy nanosecond laser irradiation, they were thermally more stable than PEG-coated gold nanorods. For the above reasons it was found that gold-silica core shell nanorods were very effective as a stable photoacoustic contrast agent. From this exercise it was established that gold-silica core-shell nanorods are promising materials for molecular photoacoustic imaging and image-guided molecular photothermal therapy.

2.5 Fluorescent Gold Nanoclusters in Biological Labeling:

In this exercise, the synthesis of water-soluble fluorescent gold nanoclusters (AuNC) capped with dihydrolipoic acid (DHLA) was carried out, and then application to biological labeling was studied.[31] The resulting AuNC-DHLA particles showed a quantum yield of around 1-3%, reduced photobleaching properties as compared with organic fluorophores, and very good colloidal stability. It was found that the uptake of AuNC-DHLA by cells did not have any acute toxicity. AuNCs could be used in place of colloidal quantum dots due to their small hydrodynamic diameter (<5 nm) and inert nature. Particularly for applications where the size and biocompatibility of the label is critical, this application has more potential. The relatively low quantum yield is the weakest point so far. From the results of exercise it can be said that fluorescent gold nanoclusters have large potential in biomedical applications.

2.6 Bioconjugated Polyelectrolyte-coated GNRs as Sensitive Optical Probes for Biological Labeling:

In this exercise, high-quality gold nanorods (GNR) were synthesized using a seed-mediated growth method.[32] Here, hydrochloric acid and silver nitrate were used to tune the (GNR) aspect ratio, and thus tune the optical properties. It was found that with the increase of concentration of both hydrochloric acid and silver nitrate, the aspect ratio of the GNRs increased. GNRs with a longitudinal surface plasmon resonance peak ranging from 600 to 1100 nm were synthesized by this method. Using two-photon microscopy, the following suitable bioconjugation GNRs were used as optical probes for two-photon imaging of cancer cells. Comparing the effect on cancer cells of unconjugated the results of GNRs the receptor-mediated uptake of bioconjugated GNRs into cancer cells was confirmed and found that uptake of the GNRs was very much reduced. From the combination of the strong two-photon photoluminescence and biocompatibility of GNRs, it was found that bioconjugated polyelectrolyte-coated GNRs could be used as sensitive optical probes for biological labeling.

2.7 GNPs as Fluorescent Probes:

Using many optical techniques like UV-visible absorption spectroscopy, fluorescence spectroscopy, fluorescence correlation spectroscopy (FCS) and fluorescence microscopy, the fluorescent properties of colloidal gold nanoparticles (GNP) were studied.[33] The sizes of the GNPs ranged from 16 to 55 nm. From this exercise it was established that GNPs showed excellent antiphotobleaching characteristics under strong light illumination. It was found that the fluorescence of GNPs was very strong, which meant that it could be easily detected by the fluorescence microscopy and FCS at the single particle level. Their intensity could be increased by increasing the excitation power. Using GNPs as fluorescent probes as a new technique for imaging of cells was developed based on the antiphotobleaching of GNPs and the easy photobleaching of cellular autofluorescence. This method was used for imaging of living HeLa cells successfully, using free GNPs and anti-epidermal growth factor receptor (EGFR)/GNPs conjugates as the fluorescent probes. In figure 2 are shown the fluorescence images of HeLa cells incubated with 38 nm GNPs under different photobleaching conditions.[33] Due to the extremely weak interference of autofluorescence from cell membranes, GNPs can be directly used in total internal reflection fluorescence microscopy. The fluorescent imaging method has high selectivity and so should have wider applications than the scattering imaging technique. GNPs as fluorescent probes have some good properties such as chemical and photostability, simple conjugation chemistry and good biocompatibility. These properties are better than those of currently-used fluorescent probes like fluorescent dyes and quantum dots. For this reason this method will provide a useful tool for cell imaging, targeted drug delivery and cancer diagnostics.

2.8 Use of GNPs to Activate Enzymes and Biocatalytic Processes:

In this exercise, another application of gold nanoparticles (GNP) in the biological field was studied. Here it was observed that by doping GNPs of appropriate size in cationic reverse micellar systems, the lipase activity was increased.[34] It was found that, in aqueous solution, GNPs deactivated and structurally deformed lipase. For this reason this enhancement in lipase activity in reverse micelles became more important. The activation effect of lipase in reverse micelles was increased with an increase in the size of the GNPs. Here it was considered that the surfactants aggregated around the gold nanoparticles where GNP acted as a polar core. As a result, large reverse micelles with an augmented interface were formed. For this reason, the activity of lipase increased. In this study it was found that a structurally deprived lipase induced by GNPs had its activity increased in reverse micelles. From this exercise it could be concluded that these unique properties GNPs could be used to activate enzymes and biocatalytic processes.

2.9 Multifunctional Gold Nanoparticles:

In this exercise a multifunctional nanoparticle was prepared by conjugating GNPs and ONT (oligonucleotide: a short nucleic acid polymer) hybridized, PSMA-specific aptamer.[35] This formed a GCrich duplex which acted as a loading site of chemotherapeutic agent, doxorubicin (Dox), which is used as a drug in cancer chemotherapy. This multifunctional nanoparticle can be used in combined prostate cancer imaging by CT and anticancer therapy. Since PSMA aptamer-conjugated GNPs were able to bind the target prostate cancer cells that overexpressed PSMA, antigen target specific binding could be visualized by silver staining and clinical CT instrumentation. Also it was observed that, using the Doxloaded aptamer-conjugated GNPs, the target cancer cells were killed more effectively than non-target cells. Thus it gives a pathway of target-specific drug delivery using the multifunctional GNP. Figure 3 shows the fluorescence spectra of a Dox solution containing different amounts of aptamer.[35] Using this concept of conjugate GNP, it can also be applied in other disease-specific aptamers and imaging nanoprobes.

2.10 Use of Gold Nanorods Applied in Molecularly-targeted Photodiagnostics and Therapy:

In this exercise it was presented that as gold nanorods have a strong absorption and scattering of near-infrared light, they were used as optically-active dual imaging therapy agents.[36] Since the molecular targeting of overexpressed EGFR on the malignant cell surface is done by using surface plasmon resonance absorption spectroscopy and light scattering imaging, the HSC (Hematopoietic stem cell) and HOC malignant cells are easily distinguished from HaCaT nonmalignant cells. Using anti-EGFR antibody-conjugated Au nanorods, a near-infrared cw Ti:sapphire laser at 800 nm with different laser power energies cause photothermal destruction among malignant and nonmalignant cells after exposure of these cells. It was found that by increasing the uptake of the nanorods by the two malignant cells, the energy needed to cause destruction of these cells was reduced to about half of that required to kill the nonmalignant cells. Thus gold nanorods are biocompatible optically-active absorbers and scatterers that may potentially be applied in molecularly targeted photodiagnostics and therapy.

2.11 Development of Gold Nanoparticles-assisted Colorimetric Assay for Detection of Cancer Cell:

To enable earlier detection of cancer, which is currently very time-consuming and expensive, the new method of colorimetric assay for the direct detection of diseased cells was developed. This is discussed in this exercise.[37] The assay was developed by using aptamer-conjugated gold nanoparticles (GNP) combined with the selectivity and affinity of aptamers. For the spectroscopic advantages of gold nanoparticles, the cancer cells were detected sensitively. It was found that the samples with the target cells showed a distinct color change, while on the other hand, non-target samples did not show any color change. Another quality of the assay was that it showed excellent sensitivity for both the naked eye as well as absorbance measurements by sophisticated instruments. The assay could also differentiate between different types of target. It can control cells which are based on the aptamer used in the assay. For this reason it can be concluded that the assay using GNP has a wide applicability for diseased cell detection. Considering the above properties, it can be said that aptamer-conjugated gold nanoparticles could be used as a powerful tool for care diagnostics.

2.12 Using Gold Nanoparticles in Drug-delivery of Platinum-based Anticancer Drugs:

In this study, a drug-delivery strategy for platinum-based anticancer drugs coated on gold nanoparticles was developed.[38] It established that a large family of gold-platinum drug nanoparticles could be produced and, as a result, it could tune and improve the technology. Here, different sized gold particles (range between 20-100 nm) were synthesized, onto which the drugs were attached. The drugs were of different lengths and structure, as well as different terminal end groups such as dicarboxylates, thiols etc. The use of different drugs like cisplatin and multinuclear drugs such as BBR3464 could be attached to the drug nanoparticles. Also, it could be attached with active targeting groups like folate and estrogen, prostate or leukemia targeting aptamers etc., lung cancer targeting peptides or B-cell lymphoma-targeting antibodies. Thus gold nanoparticles have wide application in the field of nanomedicine-based strategy. As a result they have made a different approach possible for cancer therapy and improved disease management in the treatment of cancer.

2.13 Biosensor based on Gold and Silver Nanoparticles:

In this exercise, an organic adhesion layer of gold and silver nanoparticles was immobilized onto quartz substrates.[39] These nanoparticles were functionalized with bioreceptors for use as a biosensor. A change in both the intensity and the wavelength of the particle resonances as well as the intensity of inter-band absorptions was observed when proteins were bounded to the gold and silver particles respectively. Also, a resonance enhancement of the inter-band absorption was observed. This was the first time that quartz substrates in combination with silver and gold particles were used as biosensors. This was based on a resonance enhancement of the inter-band absorption bands. This method can also be applied to the liquid and gas phases. Thus this method of making a biosensor using gold and silver nanoparticles was found to be a promising, easy and cost-effective alternative to conventional biosensing techniques.

2.14 Using Gold Nanoparticles for Glycation Sensing:

In this study, gold nanoparticles (GNP) were used as biosensors for sensing the formation of advanced glycosylated end products (AGEs) using the plasmon resonance of GNPs.[40] The GNPs were synthesized on a protein template. The progress of the glycation formation was detected by a graded alteration of plasmon resonance (both the peak and intensity). From transmission electron microscopy it was found that there was a significant shift in the size distribution of GNPs which were present in glycation. The higher plasmon resonance was observed when formation of GNPs was increased, which in turn resulted in a larger number of smaller particles. Using infrared (IR) spectroscopy and circular dichroism (CD) studies, the binding of the protein with the GNP was studied. From CD studies, the emergence of β-structure and loss of α-helix were confirmed. From the IR data it was found that glycation induced alterations in the amide I region. Hence the formation of the sensor for formation of AGEs operated by direct or indirect conjugation with amino groups. Now glycation and AGE formation are responsible for treatment of diabetes-related diseases. Thus this approach could be used for a simple colorimetric assay for the AGEs.

Reference

1. Faraday, M. The Bakerian Lecture: Experimental Relations of Gold (and Other Metals) to Light. Phil. Trans. 147, (1857): 145-181.

2. Kreibig, U. Electronic properties of small silver particles: the optical constants and their temperature dependence. J. Phys. F: Met. Phys. 4, (1974): 999-1014

3. Grandqvist C. G. Buhrman, R. A. Ultrafine metal nanoparticles. J. Appl. Phys. 47, (1976): 2200(1-20).

4. Birrmger, R. Gleiter, H. Klein, H. P. Marquardt, Nanocrystalline materials an approach to a novel solid structure with gas-like disorder?. P. Phys. Lett. A 102, (1984): 365-369.

5. Linderoth, S. Morup, S. Ultrasmall iron particles prepared by use of sodium amalgam. J. Appl. Phys. 67, (1990): 4496(1-3).

6. Roy, R. A. Roy, R. Diphasic xerogels: I. Ceramic metal composites. Mater. Res. Bull. 19, (1984): 169-177.

7. Chakravorty, D. Memory switching in ion-exchanged oxide glasses. Appl. Phys. Lett. 24, (1974): 62(1-2).

8. Chakravorty, D. Shuttleworth A. Gaskell, P. H. Microstructural studies of glass-metal composites produced by ion-exchange and hydrogen treatments. J Mater. Sci. 10, (1975): 799-808.

9. Roy, B. Chakravorty, D. Electrical conductance of silver nanoparticles grown in glass-ceramic. J. Phys.: Condens. Matter 2, (1990): 9323-9334.

10. Roy, S. Chakravorty, D. Silver electrodeposits in ion-exchanged oxide glasses. Phys. Rev. B. 47, (1993): 3089-3096.

11. Banerjee, S. Chakravorty, D. Optical absorption of composites of nanocrystalline silver prepared by electrodeposition. Appl. Phys. Lett. 72, (1998): 1027(1-3).

12. Rao, C. N. R. Blackwell Scientific Publications Ltd., Oxford pp. 217-235.

13. Starowicz, M. Stypuła, B. Banaś, J. Electrochemical synthesis of silver nanoparticles. Electrochemistry Communications 8, (2006): 227-230.

14. Yu, Y. Y. Chang, S.S. Lee, C. L. Wang, C. R. C. Gold Nanorods: Electrochemical Synthesis and Optical Properties. J. Phys. Chem. B 101,

(1997): 6661-6664.

15. Zeng, J. Zheng, Y. Rycenga, M. Tao, J. Li, Z.Y. Zhang, Q. Zhu, Y. Xia Y. Controlling the Shapes of Silver Nanocrystals with Different Capping Agents. J. AM. Chem. Soc. 132, (2010): 8552-8553.

16. Jana, N. R. Gearheart, L. Murphy C. J. Wet Chemical Synthesis of High Aspect Ratio Cylindrical Gold Nanorods. J. Phys. Chem. B 105, (2001): 4065-4067

17. Nikoobakht, B. El-Sayed M. A. Preparation and Growth Mechanism of Gold Nanorods (NRs) Using Seed-Mediated Growth Method. Chem. Mater. 15, (2003): 1957-1962.

18. Reddy, A. S. Chen, C. Y. Chen, C. C. Jean, J. S. Chen, H. R. Tseng, M. J. Fan, C.W. Wang, J. C. Biological Synthesis of Gold and Silver Nanoparticles Mediated by the Bacteria Bacillus Subtilis. Journal of Nanoscience and Nanotechnology 10, (2010): 6567-6574.

19. Kasthuri, J. Veerapandian, S. Rajendiran, N. Biological synthesis of silver and gold nanoparticles using apiin as reducing agent. Colloids and Surfaces B: Biointerfaces 68, (2009): 55-60.

20. Kasthuri, J. Kathiravan K. Rajendiran N. Phyllanthin-assisted biosynthesis of silver and gold nanoparticles: a novel biological approach. J Nanopart Res 11, (2009): 1075-1085.

21. Mukherjee, P. Ahmad, A. Mandal, D. Senapati, S.. Sainkar, S. R Khan, M. I. Parishcha, R. Ajaykumar, P. V. Alam, M. Kumar, R. Sastry M. Fungus-Mediated Synthesis of Silver Nanoparticles and Their Immobilization in the Mycelial Matrix: A Novel Biological Approach to Nanoparticle SynthesisNano Lett., 1, (2001): 515-519.

22. Song, J. Y. Kim, B. S. Rapid biological synthesis of silver nanoparticles using plant leaf extracts. Bioprocess Biosyst Eng 32, (2009): 79-84.

23. Raveendran, P. Fu, J. Wallen, S. L. Completely "Green" Synthesis and Stabilization of Metal Nanoparticles. J. AM. Chem. Soc. 125, (2003): 13940-13941.

24. Singh, A. K. Talat, M. D. Singh, P. Srivastava O. N. Biosynthesis of gold and silver nanoparticles by natural precursor clove and their functionalization with amine group. J Nanopart Res 12, (2010): 1667-1675.

25. Wu, Q. Cao, H. Luan, Q. Zhang, J. Wang, Z. Warner, J. H. Watt, A. A. R. Biomolecule-Assisted Synthesis of Water-Soluble Silver Nanoparticles and Their Biomedical Applications. Inorganic Chemistry 47, (2008): 5882-5888.

26. Shankar, S. S. Rai, A. Ahmad, A. Sastry M. Rapid synthesis of Au, Ag,

and bimetallic Au core–Ag shell nanoparticles using Neem (Azadirachta indica) leaf broth. Journal of Colloid and Interface Science 275, (2004): 496-502.

27. Eck, W. Craig, G. Sigdel, A. Ritter, G. Old, L. J. Tang, L. Brennan, M. F. Allen, P. J. Mason, M. D. PEGylated Gold Nanoparticles Conjugated to Monoclonal F19 Antibodies as Targeted Labeling Agentsfor Human Pancreatic Carcinoma Tissue. ACSNano 2, (2008): 2263-2272.

28. Cognet, L. Tardin, C. Boyer, D. Choquet, D. Tamarat, P. Lounis B. Single metallic nanoparticle imaging for proteindetection in cells. PNAS 100, (2003): 11350-11355.

29. Schrand, A.M. Braydich-Stolle, L. K. Schlager, J. J. Dai, Hussain L. S. M. Can silver nanoparticles be useful as potential biological labels?Nanotechnology 19, (2008): 235104(1-13).

30. Chen, Y. S. Frey, W. Kim, S. Homan, K. Kruizinga, P. Sokolov, K. Emelianov S. Enhanced thermal stability of silica-coated gold nanorods for photoacoustic imaging and image-guided therapy. Optics Express 18, (2010): 8868-8878.

31. Lin, C.A. J. Yang, T. Y. Lee, C. H. Huang, S. H. Sperling, R. A. Zanella, M. Li, J. K. Shen, J. L. Wang, H. H. Yeh, H. I. Parak, W. J. Chang, W. H. Synthesis, Characterization, and Bioconjugation of Fluorescent Gold Nanoclusters toward Biological Labeling Applications. ACSNano 3, (2009): 395-401.

32. Zhu, J. Yong, K. T. Roy, I. Hu, R. Ding, H. Zhao, L. Swihart, M. T. He, G. S. Cui, Y. Prasad, P. N. Additive controlled synthesis of goldnanorods (GNRs) for two-photonluminescence imaging of cancer cells. Nanotechnology 21, (2010): 285106(1-8).

33. He, H. Xie, C. J. Ren Nonbleaching Fluorescence of Gold Nanoparticles and Its Applications in Cancer Cell Imaging. Anal. Chem 80, (2008): 5951-5957.

34. Maiti, S. Das, D. Shome, A. Das, P. K. Influence of Gold Nanoparticles of Varying Size in Improving the LipaseActivity within Cationic Reverse Micelles. Chem. Eur. J. 16, (2010): 1941-50.

35. Kim, D. Jeong, Y. Y. Jon S. A Drug-Loaded Aptamer_Gold Nanoparticle Bioconjugate for Combined CT Imaging and Therapy of Prostate Cancer. ACSNano 4, (2010): 3689-3696.

36. Huang, X. El-Sayed, I. H. Qian, W. El-Sayed M. A. Cancer Cell Imaging and Photothermal Therapy in the Near-Infrared Region by Using Gold Nanorods. J.Am.Chem.Soc. 128, (2006): 2115-2120.

37. Medley, C. D. Smith, J. E. Tang, Z. Wu, Y. Bamrungsap, S. Tan, W. Gold

Nanoparticle- Based Colorimetric Assay for the Direct Detection of Cancerous CellsAnal. Chem. 80, (2008): 1067-1072.

38. Brown, S. D. Nativo, P. Smith, J. A. Stirling, D. Edwards, P. R. Venugopal, B. Flint, D. J. Plumb, J. A. Graham, D. Wheate, N. J. Gold Nanoparticles for the Improved Anticancer Drug Delivery of the Active Component of OxaliplatinJ. Am. Chem. Soc. 132, (2010): 4678-4684.

39. Frederix, F. Friedt, J. M. Choi, K. H. Laureyn, W. Campitelli, A. Mondelaers, D. Maes, G. Borghs, G. Biosensing Based on Light Absorption of Nanoscaled Gold and Silver Particles. Anal. Chem. 75, (2003): 6894-6900.

40. Ghosh Moulick, R. Bhattacharya, J. Mitra, C. K. Basak, S. Dasgupta A. K. Protein seeding of gold nanoparticles and mechanism of glycation sensing. Nanomedicine 3, (2007): 208-214.

Chapter 3
Biological Implications of Metallic Nanoparticles

Figure 1: (A) TEM and (B) HRTEM images of the spherical, single-crystal Ag seeds. (C) TEM and (D) HRTEM images of Ag octahedrons obtained by adding 1.8 mL of AgNO3 into 2.1 mL of the Ag seeds, together with 0.1 mL of 40 mM AA and 0.8 mL of 40 mM Na3CA. (E) TEM and (F) HRTEM images of Ag nanocubes and nanobars prepared under the same conditions as those for (C) except that 0.8 mL of 112 mM PVP (in terms of the repeating unit) was added instead of Na3CA. The scale bars in the insets correspond to 20 nm. (Reference: "Reprinted with permission from {J. ZENG, Y.

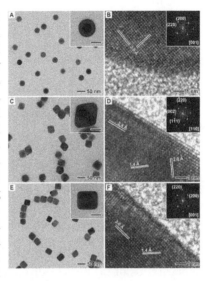

ZHENG, M. RYCENGA, J. TAO, Z.Y. LI, Q. ZHANG, Y. ZHU, and Y. XIA, J. AM. CHEM. SOC. 132, 8552 (2010)}. Copyright {2010} American Chemical society")

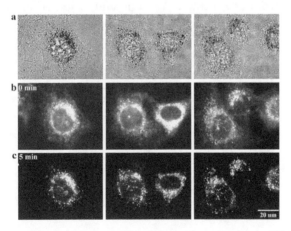

Figure 2. Bright-field images (a) and fluorescence images of HeLa cells incubated with 38 nm GNPs (b) before and (c) after 5 min photobleaching. The other two images in the same row for each imaging condition are shown to test reproducibility. Exposure time: 5 s. (Reference: "Reprinted with permission from {H. HE, C. XIE and J. REN, ANAL. CHEM 80, 5951 (2008)}. Copyright {2008} American Chemical society").

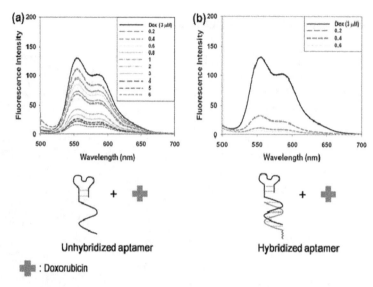

Figure 3: Fluorescence spectra of a Dox solution (3 _M) with increasing molar ratios of (a) unhybridized aptamer (from top to bottom: 0, 0.2, 0.4, 0.6, 0.8, 1, 2, 3, 4, 5, and 6 equiv) and (b) hybridized aptamer (from top to bottom: 0, 0.2, 0.4, and 0.6 equiv). (Reference: "Reprinted with permission from { D. Kim, Y. Y. Jeong and S. Jon ACSNano 4, 3689 (2010)}. Copyright {2010} American Chemical society")

Chapter 4

NON-METALLIC NANOPARTICLES & THEIR BIOLOGICAL IMPLICATIONS

With Special Reference to Carbon Nanotubes, Selenium and Sulfur Nanoparticles

Samrat Roy Choudhury and Arunava Goswami

In material science parlance, "nanoparticle" is a generic term used for nanosized metallic, metal oxide, semiconductor, magnetic and nonmetallic elements or compounds. The preparation, characterization and biological activities of nanoforms of non-metals, however, are not as widely known as their metallic nanoparticle equivalents. This can be logically correlated with the poor or insignificant properties of non-metals relative to metals in terms of their electro-thermal conductivity, electro-negativity or physicochemical stability. Therefore, in general, nonmetallic nanoparticles were not expected to demonstrate any striking difference in their physicochemical and biological properties than that of their bulk counterpart. Nonetheless, owing to extensive research in the last few years, carbon nanotubes, selenium and sulfur nanoparticles have emerged as the most promising non-metallic nanoforms in respect to their biofunctionalization.

In this chapter, primarily, we summarize the structural and functional attributes of carbon nanotube, selenium and sulfur nanoparticles. In addition, we overview the methods that have been currently developed, physicochemical characterizations and biological implications of the aforementioned nanoforms.

1. An Overview of Carbon Nanotubes (CNTs)

Hollow tubes of carbon with a diameter below 100 nanometer were reported at various times prior to the discovery of carbon nanotubes by Sumio Iijima in 1991 [1]. After development and characterization, CNTs have been and are still being used tremendously for their electrical and optical properties [2, 3]. CNTs are basically composed of concentric cylinders of graphene (allotropes of carbon). Depending upon the number of such concentric cylinders, they are categorized into single-walled carbon nanotubes (SWCNTs), double-walled carbon nanotubes (DWCNTs) or multi-walled carbon nanotubes (MWCNTs). The hemispherical wrapping ends of SWCNTs determine their electronic properties [4, 5], while the hexagonal lattice vector classifies their constructions as armchair (n,n), zigzag (n,0) and chiral (m,n) (Fig. 1). DWCNTs are coaxial structures with bi-layered graphitic sheets, and are claimed to be better electro-conductive agents than SWCNTs [6]. MWCNTs have more than two concentric cylinders, with the outer diameter between 2 and 100 nm and the inner diameter about 1-3 nm [7]

Four principal methodologies have been developed for the successful preparation of carbon nanotubes, namely electric arc discharge method, laser ablation method, chemical vapor deposition method and high pressure carbon monoxide disproportionation (HiPco) method [8].

1.1 Electric Arc Discharge

An electric arc is an electrical breakdown of a gas, which produces an ongoing plasma discharge resulting from a current flowing through normally nonconductive media, such as air. A synonym for this phenomenon is arc discharge. An arc discharge is characterized by a lower voltage than a glow discharge, and relies on thermionic emission of electrons from the electrodes supporting the arc. The arc occurs in the gas-filled space between two conductive electrodes (often made of tungsten or carbon) and results in a very high temperature capable of melting or vaporizing most materials [9]. An electric arc is a continuous discharge, while a similar electric spark discharge is momentary. An electric arc may occur either in direct current circuits or in alternating current circuits. In the latter case, the arc may re-strike on each half-cycle of the current. An electric arc differs from a glow discharge in that the current density is quite high, and the voltage drop within the arc is low; at the cathode the current density may be as high as one mega-ampere per square centimeter.

1.2 Laser Ablation Method

Laser ablation is the process of removing material from a solid (or occasionally liquid) surface by irradiating it with a laser beam. At low laser flux, the material is heated by the absorbed laser energy to be evaporated or sublimed. At high laser flux, the material is typically converted to plasma. Usually, laser ablation refers to removing material with a pulsed laser, but it is possible to ablate material with a continuous-wave laser beam if the laser intensity is high enough. The depth over which the laser energy gets absorbed is proportional to the amount of material removed by a single laser pulse, depending upon the optical properties and the laser wavelength of the material. Laser pulses can vary over a very wide range of duration (milliseconds to femtoseconds) and fluxes, and can be precisely controlled. This makes laser ablation very valuable for both research and industrial applications [9].

1.3 Chemical Vapor Deposition (CVD)

Chemical vapor deposition (CVD) is a chemical process used to produce high-purity, high-performance solid materials. The process is often used in the semiconductor industry to produce thin films. In a typical CVD process, the wafer (substrate) is exposed to one or more volatile precursors, which react and/or decompose on the substrate surface to produce the desired deposit [10, 11]. Frequently, volatile by-products are also produced, which are removed by gas flow through the reaction chamber.

1.4 High-pressure Carbon monoxide Disproportionation Method (HiPco)

The following describes the large-scale production (10 g/day) of high-purity carbon single-walled nanotubes (SWNTs) using a gas-phase chemical-vapor-deposition (HiPco process) [12]. SWNTs grow in high-pressure (30–50 atm), high-temperature (900–1100°C) flowing CO over catalytic clusters of iron. The clusters are formed *in situ*: Fe is added to the gas flow in the form of Fe $(CO)_5$. Upon heating, the Fe $(CO)_5$ decomposes and the iron atoms condense into clusters. These clusters serve as catalytic particles upon which SWNT nucleate and grow (in the gas phase) via CO disproportionation: $CO+CO \Rightarrow CO_2+C$ (SWNT). SWNT material of up to 97 mol percent purity has been produced at rates of up to 450 mg/h. The HiPco process has been studied and optimized with respect to a number of process parameters including temperature, pressure, and catalyst concentration. The behavior of the SWNT yield with respect to various parameters sheds light on the processes that currently limit SWNT production, and suggests ways that the production rate can be increased still further.

1.5 Biological Implications of CNTs

The surface of CNTs can be modulated for biological functionalities after conjugation with different biologically active organic moieties. A number of lipids, proteins including enzymes, peptides and nucleic acids can be non-covalently conjugated with CNTs and potentially used as biological sensors or markers. Quasi 1-D structure and excellent electronic properties ensures improved miniaturization prospects of CNT biosensor devices [13-15].

Chemically modified carbon nanotubes can also be used as substrates for cultured neurons. The morphological features of neurons that directly reflect their potential capability in synaptic transmission are characterized [16]. The chemical properties of carbon nanotubes are systematically varied by attaching different functional groups that confer known characteristics to the substrate. By manipulating the charge carried by functionalized carbon nanotubes we are able to control the outgrowth and branching pattern of neuronal processes.

Biological systems are known to be highly transparent to 700-1,100 nm in near-infrared (NIR) light. It is shown here that the strong optical absorbance of single-walled carbon nanotubes (SWNTs) in this special spectral window, an intrinsic property of SWNTs, can be used for optical stimulation of nanotubes inside living cells to afford multifunctional nanotube biological transporters. For oligonucleotides transported inside living cells by nanotubes, the oligos can translocate into the cell nucleus upon endosomal rupture triggered by NIR laser pulses. Continuous NIR radiation can cause cell death because of excessive local heating of SWNT *in vitro*. Selective cancer cell destruction can be achieved by functionalization of SWNT with a folate moiety, selective internalization of SWNTs inside cells labeled with folate receptor tumor markers, and NIR-triggered cell death, without harming receptor-free normal cells. Thus the transporting capabilities of carbon nanotubes combined with suitable functionalization chemistry and their intrinsic optical properties can lead to new classes of novel nanomaterials for drug delivery and cancer therapy [17].

2. An Overview of Selenium Nanoparticles

Elemental selenium is a rarely-occurring nonmetal and more commonly exists as inorganic selenide, selenate or selenite. Although high doses of selenium are considered to be potentially toxic to living organisms, a low dose of selenium supplement is essential for our body. Deficiency of selenium may induce skin cancer or tuberculosis. Synthesis of nanoform of selenium is a recent development. Many researchers have developed a number of methodologies for the preparation of selenium nanoparticles. Some of the promising techniques for the aforementioned purpose are as follows:

2.1 Selenium Nanoparticles on Cellulose Nanocrystals

Selenium nanoparticles of 10–20 nm in diameter have been prepared using cellulose nanocrystal (CNXL) as a reducing and structure-directing agent under hydrothermal conditions [18]. Na_2SeO_3 was reduced to form elemental selenium nanoparticles under hydrothermal conditions. During the hydrothermal process (120–160 °C), CNXL rods were mainly maintained and selenium nanoparticles were interfacially bound to the CNXL surface. The reaction temperature affects the sizes of interfacially-bound selenium nanoparticles (Fig 2 and 3).

2.2 Selenium Nanoparticles from Sodium selonosulfate

A simple wet chemical method has been developed to synthesize glucose-stabilized selenium nanoparticles from an aqueous sodium selenosulphate precursor. Selenium nanoparticles were prepared by the reduction of aqueous sodium selenosulphate solution with freshly prepared glucose solution. The method is capable of producing selenium nanoparticles in a size range of about 20-80 nm, under ambient conditions. The glucose-stabilized selenium particles can be used as biocompatible material for biological applications [19].

2.3 Selenium Nanoparticles wrapped within Chitosan Polymer

Natural biopolymer chitosan cross-linked with glutaraldehyde was selected as a template to synthesize selenium nanoparticles because of its unusual property of combination, including excellent membrane-forming ability, high permeability towards water, good adhesion, biocompatibility and high mechanical strength. Then the resulting selenium nanoparticles were coated on glassy carbon electrode to form a stable and even film. Finally, we immobilized horseradish peroxidase (HRP) onto the selenium nanoparticle layer to develop the H_2O_2 biosensor [20].

2.4 Biosynthesis of Selenium Nanoparticles

A simple wet analytical technique (a titrimetric method) was first employed for a preliminary evaluation of the reduction potential of *Klebsiella pneumonia* (enterobacteriaceae) grown in various culture media. The reduction capability of the specific reported strain was capable of converting Se^{+4} to elemental selenium. Optimum yield of selenium nanoparticle was observed while the strains were cultured on tryptic soya broth at specific culture conditions [21]. The nanoparticles thus obtained were isolated and characterized for their purity.

2.5 Biological Implications of Selenium Nanoparticles

Selenium nanoparticles have been utilized for several advanced biophysical and biochemical purposes. Many research groups have already reported different novel aspects related to the biological applications of selenium nanoparticles which are as follows:

2.5.1 Inhibition of oxidative stress by selenium nanoparticles

Melatonin-selenium nanoparticles (MT-Se) complex was designed to determine the protective effect of the prepared complex against Bacillus Calmette–Guérin (BCG)/lipopolysaccharide (LPS)-induced hepatic injury in mice. In BCG/LPS-induced hepatic injury model, MT-Se administered at doses of 5, 10, or 20 mg/kg to BCG/LPS-treated mice for 10 days significantly reduced the increase in plasma aminotransferase, and reduced the severe extent of hepatic cell damage and the immigration of inflammatory cells. The MT-Se particles also attenuated the increase in the content of thiobarbituric acid-reactive substances and enhanced the decrease in reduced activities of superoxide dismutase and glutathione peroxidase (GPx). However, treatment with MT-Se suppressed the increase in nitric oxide levels both in plasma and liver tissue. Furthermore, supplementation with MT-Se at the dose of 10 mg/kg (composed of 9.9 mg/kg melatonin and 0.1 mg/kg selenium) had greater capability to protect against hepatocellular damage than a similar dose of melatonin (10 mg/kg) or selenium (0.1 mg/kg) alone. This effect may relate to its higher antioxidant efficacy in decreasing lipid peroxidation and increasing GPx activity. These results suggest that the mode of MT-Se hepatic protective action is, at least in part, related to its antioxidant properties [22]

2.5.2 Selenium nanoparticles-mediated induction for mitochondrial apoptosis

Low-toxic selenium nanoparticles are used for the treatment of A375 human melanoma cells in a dose-dependent manner that causes cell apoptosis, as indicated by DNA fragmentation and phosphatidylserine translocation. Further investigation on intracellular mechanisms found that Se-NPs treatment triggered apoptotic cell death in A375 cells with the involvement of oxidative stress and mitochondrial dysfunction. Our results suggest that Se-NPs may be a candidate for further evaluation as a chemopreventive and chemotherapeutic agent for human cancers, especially melanoma cancer [23].

***2.5.3** Cytotoxic study with selenium nanoparticles against cancers cells*
Selenium nanoparticles (SeNPs) were prepared based on the reduction of selenious acid (H_2SeO_3), by employing sodium alginate (SA) as a template. The real-time monitoring of the drug-inducing apoptosis process of human hepatic cancer cells Bel7402 was performed with the quartz crystal microbalance (QCM) measurement. The anti-tumor effect of adriamycin (ADM) used in combination with SeNPs was investigated. It was found that both drugs were able to inhibit cell proliferation in a dose-dependent way, and the combined treatment with ADM and Se-NPs was more effective in inhibiting cell growth than each of the two drugs alone. The cytotoxic effects of drug combination were evaluated with suitable standard biochemical assays. The grades gradually changed from apparent synergism to simple addition with the drug-treatment time increasing but the drug combination with lower concentrations still exhibiting synergism after 24 hours, suggesting a potential application in cancer therapy [24].

3. An Overview of Sulfur Nanoparticles

The pungent sulfurous smell of sulfur has always tended to fascinate human beings, and has lead to the elevation of this element to an empyrean status. Biblical Pentateuch states that the destruction of the twin cities of Sodom and Gomorrah was purported to be mediated by brimstone (sulfur) and fire. However, sulfur is not only associated with death and destruction. This multivalent inorganic nonmetal has been in use since time immemorial for its purifying and beneficial properties. The Egyptians were familiar with sulfurous antiseptic cream for treating bacterial infestations, while the Chinese injected colloidal sulfur to treat rheumatoid arthritis. So from a divine, fearsome substance, across the mystical alchemists' universe to the group 16 (VIB) of the modern periodic table, the origin- and broad-spectrum implications of sulfur have been profusely explored.

Sulfur (S^0) and sulfur-rich compounds bear an interesting interaction with biological systems. Diverse floral and microbial communities assimilate a reduced form of S^0 through direct sulfhydrylation pathway [26, 27], convert them into sulfide and finally reduce it to thiol (the lowest oxidation state of sulfur). Formation and incorporation of thiol into methionine is quintessential for the initiation of protein synthesis machinery in all the living organisms. It is interesting to note that while sulfur plays a crucial role in protein synthesis at a basal quantity, a high concentration of sulfur is relatively toxic to microorganisms due to its ability to denature certain proteins and bind to metal-centered enzymes [28-30]. Moreover, diverse plant families are found to accumulate S^0 or sulfur-rich organic compounds as phytoalexin in order to combat pathogenic invasions [31-34]. This non-systemic and contact pesticide is also widely used in agro-medical fields, either in its wettable form, dust form, colloidal form or in combination with other synthetic and organic pesticides [35, 36]. Various forms of S^0 have been used to control brown rot in peaches, powdery mildew in apples, gooseberries, grapes, strawberries etc., scab in roses and certain smut and rust fungi of hops and ornamental plants. Earlier studies have suggested that S^0 is also efficient against many bacterial species of agricultural and medical importance [30, 37]. Antimicrobial efficacy of S^0 hence has been exploited for decades for treating bacterial and fungal infections [30, 35, 36, 37]. However, genetic modifications and concomitant acquired resistance in target pathogens depreciates S^0 as an effective biocide. At the same time, due to its large volume requirement and high cost, it has lost its worldwide popularity among farmers and agro-chemical industries.

Several methodologies proposed for the preparation of sulfur nanoparticles are as follows:

3.1 Liquid Synthesis Method

A novel liquid-synthesis process of preparing monoclinic sulfur nanoparticles in the presence of surfactant PEG-400 is developed by using sublimed sulfur powders and ammonium sulfide as an S^{2-} precursor, and acetic acid as a precipitating agent. PEG-400 can be used to form a molecule chain and surround the reactants to prevent the product molecules from aggregating, in order to obtain nanoparticles in small diameters. The size of sulfur nanocrystrals is well controlled at 40–60 nm in this process [38, 39].

3.2 Cysteine Modification Method

A novel sulfur form, insect-like nanoelemental sulfur (nano-S^0), was prepared by adding cysteine to sulfur-ethanol solution. The modification and modulation of cysteine on elemental sulfur (S^0) nanoparticles in liquid phase was studied. The produced nanosulfur was characterized by TEM, RRS, IR and Raman Spectra. The results show that the cysteine could modulate the S^0 nanoparticles in the aspects of size, morphology and stabilization. The produced cysteinenano-S^0 particles in the sol were insect-like. The modulation and stability of cystine on S^0 nanoparticles were considered to have taken place be by electrosteric means and by the formation of ligands [40].

3.3 Wate- in-Oil Microemulsion Technique

Nanosized monoclinic sulfur particles have been successfully prepared via the chemical reaction between sodium polysulfide and hydrochloric acid in a reverse microemulsions system, with theolin, butanol, and a mixture of Span80 and Tween80 (weight ratio 8: 1) as the oil phase, cosurfactant and surfactant, respectively. Transparent microemulsions were obtained by mixing the oil phase, surfactant, co-surfactant, and the aqueous phase in appropriate proportion, using an emulsification machine at room temperature [41].

3.4 Biological Implications of Sulfur Nanoparticles

Surface-modified sulfur nanoparticles (Fig. 4) have been used as an effective antifungal agent against both sulfur-resistant (*Aspergillus niger*) and sulfur-susceptible fungus (*Fusarium oxysporum*) [42, 43]. The rationale of the study was to revalidate the antifungal role of sulfur in agricultural fields.

The antibacterial role of sulfur nanoparticles has also been reported [44] but not explored sufficiently, and needs great attention in this regard.

Reference

1. Iijima S (1991) Helical microtubules of graphitic carbon. Nature 354, 56-58 (07 November 1991); doi:10.1038/354056a0

2. Lu X., Chen Z (2005) Curved Pi-Conjugation, Aromaticity, and the Related Chemistry of Small Fullerenes (<C60) and Single-Walled Carbon Nanotubes". Chemical Reviews 105 (10): 3643–3696. doi:10.1021/cr030093d. PMID 16218563

3. Wang X, Li Q, Xie J, Jin Z, Wang J, Li Y, Jiang K, Fan S (2009). "Fabrication of Ultralong and Electrically Uniform Single-Walled Carbon Nanotubes on Clean Substrates". Nano Letters 9 (9): 3137–3141. doi:10.1021/nl901260b. PMID 19650638

4. Maohui Ge et al. Appl. Phys. Lett. 65 (18), 2284 (1994).

5. Klaus Sattler Carbon 33(7), 915 (1995)

6. Pfeiffer R, Pichler T, Ahm Kim Y, Kuzmany H (2008) Topics in Applied Physics, Volume 111/2008, 495-530, DOI: 10.1007/978-3-540-72865-8_16

7. Beckyarova E, Haddon RC, Parpura V (2005) Biofunctionalization of carbon nanotubes,

8. Book: Biofunctionalization of nanomaterials, 1st edn, Willey-Vch Verlag GmbH & Co. KGaA

9. Journet C, Bernier P (1998) Production of carbon nanotubes. Applied Physics A 67(1-9)

10. Howatson AH, An Introduction to Gas Discharges, Pergamon Press, Oxford pgs. 80-95

11. Guo T, Nikolaev P, Thess A, Colbert DT, Smalley RE (1995). "Catalytic growth of single-walled nanotubes by laser vaporization". Chem. Phys. Let. 243: 49. doi:10.1016/0009-2614(95)00825-O

12. Jaeger, Richard C. (2002). "Film Deposition". Introduction to Microelectronic Fabrication. Upper Saddle River: Prentice Hall. ISBN 0-201-44494-7.

13. Michael J. Bronikowski, Peter A. Willis, Daniel T. Colbert, K. A. Smith, and Richard E. Smalley (2001) Gas-phase production of carbon single-walled nanotubes from carbon monoxide via the HiPco process: A parametric study. Journal of Vacuum Science & Technology A ,Volume 19(4) / nanotubes/nanometer-scale science and technology.

14. Koen Besteman, Jeong-O Lee, Frank G. M. Wiertz, Hendrik A. Heering, and Cees Dekker (2003) Enzyme-Coated Carbon Nanotubes as Single-Molecule Biosensors. Nano Letters, 2003, 3 (6), pp 727–730 DOI: 10.1021/nl034139u

15. Cyrille Richard, Fabrice Balavoine, Patrick Schultz, Thomas W. Ebbesen and Charles Mioskowski (2003) Science: Vol. 300 no. 5620 pp. 775-778. DOI: 10.1126/science.1080848

16. Sandeep S. Karajanagi, Alexey A. Vertegel, Ravi S. Kane and Jonathan S. Dordick (2004) Structure and Function of Enzymes Adsorbed onto Single-Walled Carbon Nanotubes. Langmuir: 20 (26), pp 11594–11599 .DOI: 10.1021/la047994h

17. Hui Hu,Yingchun Ni,Vedrana Montana,Robert C. Haddon and Vladimir Parpura (2004) Chemically Functionalized Carbon Nanotubes as Substrates for Neuronal Growth Nano Lett 4(3): 507–511.,doi: 10.1021/ nl035193d.

18. Nadine Wong Shi Kam, Michael O'Connell , Jeffrey A. Wisdom and Hongjie Dai (2005) Carbon nanotubes as multifunctional biological transporters and near-infrared agents for selective cancer cell destruction. Published online before print August 8, 2005, doi:10.1073/ pnas.0502680102PNAS August 16, 2005 vol. 102 no. 33 11600-11605

19. Yongsoon Shin, Jade M. Blackwood, In-Tae Bae, Bruce W. Arey and Gregory J. Exarhos (2007) Synthesis and stabilization of selenium nanoparticles on cellulose nanocrystal. Material letters Volume 61, Issue 21, August 2007, Pages 4297–4300

20. Ingole AR, Thakare SR, Khati NT, Wankhade AV, Burghate DK (2010) Green synthesis of selenium nanoparticles under ambient condition. Chalcogenide Letters Vol. 7, No. 7, July 2010, p. 485–489

21. Zhang J, Zhang SY, XU JJ, Chen HY (2004) A New Method for the Synthesis of Selenium Nanoparticles and the Application to Construction of H2O2 Biosensor. Chinese Chemical Letters Vol. 15, No. 11, pp 1345-1348, 2004

22. Fesharaki PJ, Nazari P, Shakibaie M, Rezaie S, Banoee M, Abdollahi M, Shahverdi AR (2010) Biosynthesis of selenium nanoparticles using Klebsiella Pneumoniae and their recovery by a simple sterilization process. Brazilian Journal of Microbiology (2010) 41: 461-466.

23. Wang H, Wei W, Zhang SY, Shen YX, Yue L, Wang NP, XU SY (2005) Melatonin-selenium nanoparticles inhibit oxidative stress and protect against hepatic injury induced by Bacillus Calmette–Guérin/ lipopolysaccharide in mice. Journal of Pineal Research. Volume 39, Issue 2, pages 156–163

24. Chen T, Wong YS, Zheng W, Huang L (2008) Selenium nanoparticles fabricated in Undaria pinnatifida polysaccharide solutions induce mitochondria-mediated apoptosis in A375 human melanoma cells.

Colloids and Surfaces B: Biointerfaces, Volume 67, Issue 1, 15 November 2008, Pages 26-31

25. Tan L, Jia X, Zhang Y, Tang H, Yao S, Xie Q (2009) In vitro study on the individual and synergistic cytotoxicity of adriamycin and selenium nanoparticles against Bel7402 cells with a quartz crystal microbalance. Biosensors and Bioelectronics. Volume 24, Issue 7, 15 March 2009, Pages 2268-2272

26. Hunter, Christy A (2009) Application of the Quartz Crystal Microbalance to Nanomedicine. Journal of Biomedical Nanotechnology, Volume 5, Number 6, December 2009 , pp. 669-675(7)

27. Foglino M, Borne F, Bally M, Ball G, Patte JC (1995) A direct sulfhydrylation pathway is used for methionine biosynthesis in Pseudornonas aeruginosa. Microbiology 141:431-439

28. VermeiJ P, KertesZ MA (1999) Pathways of Assimilative Sulfur Metabolism in Pseudomonas putida. Journal of bacteriology 181(18):5833–5837

29. Brock MT, Madigan TD (1991) Biology of microorganisms. 6th ed. Prentice-Hall Inc., New York

30. McCallan SEA (1949) The nature of the fungicidal action of copper and sulfur. The botanical review 15:629-643

31. Libenson L, Hadley FP, Mcllroy AP, Wetzel VM, Mellon RR (1953) Antibacterial effect of elemental sulfur. Journal of infectious diseases 93:28-35

32. Williams JS, Cooper RM (2004) The oldest fungicide and newest phytoalexin a reappraisal of the fungitoxicity of elemental sulphur. Plant Pathology 53:263-279

33. Cooper RM, Williams JS (2004) Elemental sulphur as an induced antifungal substance in plant defence. Journal of Experimental Botany 55(404):1947-1953

34. Williams JS, Hall SA, Hawkesford MJ, Beale MH, Cooper RM (2002) Elemental sulfur and thiol accumulation in tomato and defense against a fungal vascular pathogen. Plant Physiology 128:150-159

35. Resende MLV, Flood J, Ramsden JD, Rowan MG, Beale MH, Cooper RM (1996) Novel phytoalexins including elemental sulfur in the resistance of cocoa (Theobroma cacao L.) to Verticillium wilt (Verticillium dahlia Kleb.). Physiol Mol Plant P 48:347-349

36. Rose RL, Roe RM, Hodgson E (1999) Pesticides, In Toxicology. Marquardt H, Schafer SG, McClellan R, Welsch F (Ed) 663-697 Academic Press, San Diego

37. Baldwin MM (1950) Sulfur in fungicides. Industrial engineering & chemistry 42(11):2227-2230

38. Lawson GB (1934) The inhibitory action of sulfur on the growth of tubercle bacilli. American review of tuberculosis 29: 650-651

39. Guo Y, Zhao J, Yang S, Yu K, Wang Z, Zhang H (2006) Preparation and characterization of monoclinic sulfur nanoparticles by water-in-oil microemulsion technique. Powder Technol 162:83-86

40. Guo Y, Deng Y, Hui Y, Zhao J, Zhang B (2005) Synthesis and characterization of sulfur nanoparticles by liquid phase precipitation method. Acta Chim Sinica 63:337-340

41. Xie XY, Zheng WJ, Bai Y, Liu J (2009) Cystine modified nano-sulfur and its spectral properties. Materials Letters 63 (2009) 1374–1376

42. Guo Y, Zhao J, Yang S, Yu K, Wang Z, Zhang H (2006) Preparation and characterization of monoclinic sulfur nanoparticles by water-in-oil microemulsions technique. Powder Technology 162 (2006) 83 – 86

43. Roy Choudhury S, Ghosh M, Mandal A, Chakravorty D, Pal M, Pradhan S, Goswami A (2011) Surface-modified sulfur nanoparticles: an effective antifungal agent against Aspergillus niger and Fusarium oxysporum. Appl Microbiol Biotechnol. 2011 Apr;90(2):733-43. Epub 2011 Feb 25.

44. Roy Choudhury,S, Nair kk, Kumar R, Gogoi R, Srivastava C, Gopal M, Subhramanyam BS, devakumar ,Goswam A (2010) AIP Conf. Proc. – October 4, 2010 – Volume 1276, pp. 154-157 International Conference on Advanced Nanomaterials and Nanotechnology (ICANN-2009); doi:10.1063/1.3504287

45. Deshpande AS, Khomane RB, Vaidya bK, Joshi RM, Harle AS Kulkarni BD (2008) Sulfur Nanoparticles Synthesis and Characterization from H2S Gas, Using Novel Biodegradable Iron Chelates in W/O Microemulsion. Nanoscale Research Letters, Volume 3, Number 6, 221 229, DOI: 10.1007/ s11671-008-9140-6

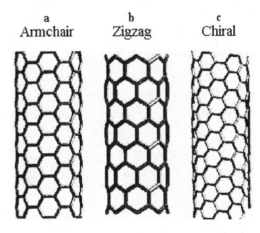

Figure 1. Different types of single walled carbon nanotube (SWNTs). Structural organization of SWNTs represented as armchair (a), zigzag (b) and chiral (c) (Reference: Scarselli, M., Castrucci, P., Crescenzi, M. D. Electronic and optoelectronic nano-devices based on carbon nanotubes. J. Phys. Cond. Matter. 24 (31), 3202, 2012).

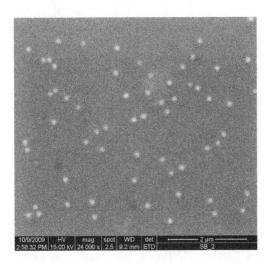

Figure.2. Scanning Electron Micrograph of monodispersed orthorhombic sulfur nanoparticles. (Reference: Roy Choudhury, S. et al. Nanosulfur: a potent fungicide against food pathogen, Aspergillus niger. AIP. Conf. Proc. 1276 (1), 154-157, 2010).

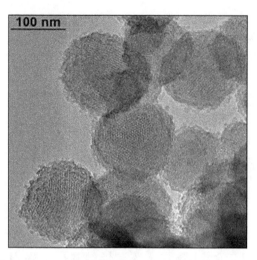

Figure 3. Transmission Electron Micrograph of selenium nanoparticles
(Resource: Goswami et al. unpublished data).

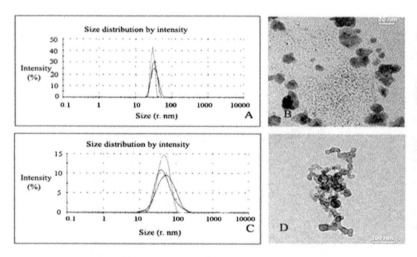

Figure 4. Dynamic Light Scattering images (A & C) and Transmission Electron Micrographs (B & D) of two different sized sulfur nanoparticles. (Reference: Roy Choudhury, S. et al. Surface modified sulfur nanoparticles: an
effective antifungal agent against Aspergillus niger and Fusarium Oxysporum.
Appl Microbiol Biotechnol. 90(2), 733-743 (2011).

Chapter 5

MAGNETIC NANOPARTICLES

Arindam Pramanik, Panchanan Pramanik

Magnetism was discovered by ancient people in lodestones, naturally created magnetized pieces of iron ore. The earliest known descriptions of magnets are from 2500 years ago from Greece, India and China [1-3]. Lodestones were the first magnetic compasses. By the twelfth and thirteenth centuries AD, magnetic compasses were used in navigation in China, Europe and elsewhere. Over the last 50 years, applications of magnetism include fields as diverse as magnetic recording media (audio tapes, computer floppy disks, hard disks), electronic motors and generators, televisions and monitors, speakers and microphones, credit and debit cards, mass transport systems like the magnetic levitation trains and magnetic resonance imaging for medical diagnostic applications.

Developments in the newly emerging areas of nanoscience and nanotechnology led to the creation of magnetic nanoparticles (MNPs). These are magnets of the size of 1-100 nm. The size depends mainly on the synthesis process and chemical structure of the final product. Synthesis is chiefly carried out by three processes, namely co-precipitation, thermal decomposition and micro emulsion [4]. Co-precipitation is a facile and convenient way to synthesize iron oxides (either Fe_3O_4 or γ-Fe_2O_3) from aqueous Fe^{2+}/Fe^{3+} salt solutions by the addition of a base under inert atmosphere at room temperature or at elevated temperature. Chlorides, sulfates or nitrates salts of iron are used.

Thermal decomposition techniques use organometallic compounds in high-boiling organic solvents containing stabilizing surfactants. In this way, monodisperse magnetic nanocrystals of small size can be synthesized. In the microemulsion technique, metallic cobalt, cobalt/platinum alloys and gold-coated cobalt/platinum nanoparticles have been synthesized in reverse micelles of cetyltrimethlyammonium bromide, using 1-butanol as the cosurfactant and octane as the oil phase.

Among the broad spectrum of nanoscale materials being investigated for biomedical use, MNPs have gained significant attention due to their intrinsic magnetic properties, which enable tracking through the radiology cornerstone, magnetic resonance (MR) imaging [5]. This class of nanoparticles includes metallic, bimetallic and iron oxide MNPs [5-6]. The latter has been widely favored because of its inoffensive toxicity profile [7-9] and reactive surface that can be readily modified with biocompatible coatings [10-12] as well as targeting, imaging and therapeutic molecules [12-15]. This flexibility has led to MNP use in magnetic separation [16], biosensor [17, 18], *in vivo* medical imaging [5, 19, 20], drug delivery [15-21], tissue repair [22] and hyperthermia [23] applications.

Currently, a number of MNPs are in early clinical trials or experimental study stages [5, 6, 8], and several formulations have been approved for clinical use for medical imaging and therapeutic applications. Notable examples include: Lumiren® for bowel imaging [8], Feridex IV® for liver and spleen imaging [23], Combidex® for lymph node metastases imaging [24], and most recently, Ferumoxytol ® for iron replacement therapy [25]. The physicochemical profiles of these MNPs provide passive targeting, but not the higher level targeting offered by bioligands. Addition of bioactive molecules to the MNP surface can increase the targeting specificity of nanoparticles [5,6,14,26,27], producing contrast agents that specifically illuminate targeted tissue, and drug carriers that do not interact with healthy tissue [5,15,16,27-30]. Development in this area represents a majority of MNP research today [Figure 1].

1. Problems arising from Nanotherapeutics

Intrinsic to the body's defense system are a series of "biological barriers" that serve to protect the body against foreign entities, including injected therapeutics and contrast agents, keeping them from reaching their intended destinations [31]. These barriers can restrict nanoparticle function by blocking their movement, causing physical changes to them, or by inducing a negative host response using biochemical signaling [32]. Upon intravascular administration, nanoparticles immediately encounter blood, a high-ionic-strength, heterogeneous solution that can induce nanoparticle agglomeration, altering their magnetic properties and inducing particle sequestration. Additionally, nanoparticles can nonspecifically interact with plasma proteins (which can trigger the adaptive immune system), extracellular matrices, and non-targeted cell surfaces while in the blood stream [33]. In each case, the nanoparticle is in danger of prematurely binding to or being taken up by cells before reaching its target tissue. In addition to coping with the vascular environment, nanoparticles must overcome various anatomical size restrictions which limit their access to target tissue (e.g. extravasation of lymph-targeting nanoparticles from the blood vessels) [31]. These size limitations are especially stringent when targeting certain organs like the brain and kidney [34]. For instance, in the brain, endothelial cells and reinforcing astrocyte cells limit levels of pinocytosis and form tight junctions between cells at the blood brain interface, yielding a structural and metabolic barrier referred to as the blood brain barrier (BBB) [35]. Here, only nanoparticles of sufficiently small sizes and appropriate physicochemical properties may pass the BBB.

2. Coating of Magnetic Particle

Unmodified magnetic nanoparticles are stable in high and low pH solutions, but use *in vivo* requires that they be coated. These surface coatings typically comprise of small organic molecules and polymers. They protect against agglomeration, provide stability and also enhance particle pharmacokinetics, endosomal release, and tailored drug loading and release actions. Chitosan, Polyethyleneimine(PEI), Polyethylene glycol(PEG), dextran and phospholipids have be employed for the purpose of successful coating of these magnetic nanoparticles [Figure 3].

2.1. Chitosan

Chitosan is a cationic, hydrophlilic polymer that is also nontoxic, biocompatible, and bioabsorbable, which has made it a favorable material for biomedical applications [36, 37]. This is primarily due to its large abundance in nature, biocompatibility and ease of functionalization. Of late, chitosan had been used in combination with MNPs [38-44].

To coat magnetic nanoparticles with chitosan, it has been found that direct, *in situ* coating is troublesome because of chitsan's poor solubility at neutral pH values which precipitates the MNPs [37]. Chitosan-coated MNPs have been produced by physically adsorbing chitosan onto oleic acid-coated nanoparticles yielding spherically shaped MNP (15 nm diameter) [43]. These MNPs were used to enhance gene transfection. In addition to its bioadsorbtive properties, chitosan possesses both amino and hydroxyl functional groups, which can be used for MNP functionalization with targeting, imaging and therapeutic agents. The cationic nature of the polymer allows complexation with genetic material, making it suitable for use as a gene delivery carrier, even when used as MNP coating. Bhattarai et al loaded chitosan-coated iron oxide nanoparticles with anionic adenovirus vectors through electrostatic interactions [42].

2.2. Polyethyleneimine (PEI)

PEI is another water-soluble positively charged polymer that can take both linear and branched forms [45]. PEI-based polymers have been used for gene delivery and for its ability to complex with DNA, to facilitate endosomal release via the "proton sponge effect," and for delivery into the nucleus [45, 46]. To improve on these properties, PEI has been integrated into MNP coatings in recent years [47-51]. Naturally, the most common application of these constructs has been for *in vitro* cell transfection with either DNA or siRNA nucleotides [47, 52]. MNP attachment has been performed by *in situ* coating [51], postsynthesis adsorption [53] and post-synthesis grafting [49, 50].

2.3. Polyethylene glycol (PEG)

Polyethylene glycol is a biocompatible linear synthetic polyether that can be prepared with a wide range of sizes and terminal functional groups. They are neutral, hydrophilic molecules in biological fluids, which helps to improve the dispersity and blood circulation time of the MNPs they are bound to [74, 54, 55, 56-58]. PEG-coated MNPs are commonly regarded as "trojan" nanoparticles because they are not readily recognized by the cells. This limits their use in imaging macrophages or other cells [59], but the same characteristic makes them ideally suited for use in target-specific cell labeling after modification with targeting ligands [60-62].

At sizes below 100 KDa, PEG polymers are usually bamphiphilic and soluble in both water and many organic solvents, including methylene chloride, ethanol, toluene, acetone and chloroform. This allows PEG assembly at the MNP surface using a variety of chemistries that require use of either aqueous or organic solvents. *In situ* coating of PEG onto MNPs precipitated under aqueous conditions [63] and via a silane group in the organic solvent, toluene [55], has already been demonstrated. Significantly, in the latter study, a hetrobifunctional PEG was prepared that could be covalently attached to the MNP surface at one end and then functionalized with targeting ligands, imaging reporter molecules or therapeutic agents at the other end [77, 78, 82, 60, 61, 64].

2.4 Dextran

Conventional dextran coatings are based on hydrogen bonding, making the polymer susceptible to detachment, but the polymers have been cross-linked after MNP attachment using epichlohydrin and ammonia, forming a Cross-Linked Superparamagnetic Iron Oxide (CLIO) [65]. CLIOs have become a versatile platform that has demonstrated a high circulation half-life in blood with no acute toxicity [66]. However, due to the use of epichlohydrin and an inability to degrade and clear from the body, their use in a clinical setting is limited [6].

2.5 Liposomes and Micelles

Liposomes and micelles, spherical aggregates of amphiphilic molecules, can be used to coat MNPs in two ways: post-synthesis incorporation, or by synthesizing MNPs directly within their open core. In the first case, water-soluble MNPs have been confined to the aqueous center of the liposome [67, 68], or alternatively, hydrophobic MNPs can be coated with micelles around the structure [69-71]. In the second case, MNPs can be precipitated in the liposomal core, which yield highly uniform nanoparticles with sizes as small as 15 nm diameter [70]. MNP-coating with either liposomal or micellular structures provide MNPs with advantages, especially when using the construct in drug delivery applications, including: simple and easy surface modification, convenient encapsulation of pharmaceuticals inside the amphiphilic substructures, and sequestration and protection of pharmaceuticals from the body until degraded in target cells [72].

3. Targeting of Magnetic Nanoparticles for Diagnosis and Detection

A number of MNP systems have implemented targeting-ligands into their design with varying success, including: small organic molecules [74, 76, 77], peptides [78-81], proteins [82], antibodies [83-85] and aptamers [86-88]. In addition to the type of ligand used, active targeting is affected by targeting molecule density and by the size and shape of the nanoparticle.

3.1 Physical targeting

Physical targeting can be achieved via the guidance of magnetic nanoparticles through an external magnetic field. Site-specific delivery of chemotherapeutic agents can be significantly improved by magnetic drug targeting (MDT). The process in general involves:

(i) the attachment of a cytotoxic drug to a biocompatible MNP carrier (i.e. magnetic targeted carrier or MTC);

(ii) intravenous injection of these MTCs in the form of a colloidal suspension;

(iii) application of a magnetic field gradient to direct the MTC to the pathological site and (iv) release of the therapeutic agent from the MTC. Although seemingly straightforward, there are many variables that complicate the execution of this technique. The effectiveness of this method depends upon a number of parameters such as the physicochemical properties of the drug-loaded MNP, field strength and geometry, depth of the target tissue, rate of blood flow and vascular supply. Early clinical trials of colloidal iron oxide MTCs loaded with epirubicin and directed toward solid tumors have demonstrated successful accumulation in the target site in about half the patients in this study [89]. A few years later, Alexiou et al demonstrated the successful *in vivo* delivery of MCT composed of starch-coated UULTRA MNPs loaded with mitoxantrone into VX2-squamous cells carcinomas on the hind limbs of New Zealand White Rabbits [90]. Although these MTC s were well-tolerated *in vivo,* some problems associated with this technique include the possibility of embolization within the blood vessels, limited field penetration of commercial magnets, control of drug diffusion after release from the MTC, and toxic responses to the MTCs. To address some of these issues and develop a theoretical basis for this technique, Grief and Richardson have created a mathematical model incorporating the effects of hydrodynamics within blood vessels, particle volumes, magnetic field strength, and even the effects of cells within the plasma. In this study the authors concluded that MDT could only be used effectively for targets close to the surface of the body [91].

3.2. Passive Targeting

3.2.1. Enhanced permeability and retention (EPR)
The enhanced permeation and retention effect that exploits the gaps in the leaky capillaries associated with cancers are common techniques used for passive targeting [92].

3.2.2. Size-dependant distribution of MNPs in tissues
Nanoparticles smaller than the fenestrations can enter the interstitium

and be entrapped in the tumor. Because of their overall size (over 40 nm in diameter), MNPs are efficiently accumulated in MPS organs. 80% of the injected dose accumulates in liver and 5–10% in the spleen, with plasma half-life lower than 10 minutes. Therefore, MNPs decrease liver and spleen signal within several minutes of intravenous administration. Malignant tumors or metastases, lacking a substantial number of Kupffer cells (liver macrophages), appear as hyperintense lesions contrasted against the hypointense liver on $T2$-weighted sequences. MNPs are routinely administered by drip infusion over a period of 30 minutes rather than with bolus injections. ULTRA MNPs exhibit an overall hydrodynamic diameter lower than 40 nm. Their smaller size enables them to act as stealth particles. Their plasma half-life is higher than 2 hours, and they therefore remain in the blood long enough to act as a blood-pool agents for MR angiography (MRA). Some particles leak to the interstitium, where they are cleared by the macrophages of the lymphatic system, or drained via the lymphatic system and subsequently accumulated in the lymph nodes. Therefore they allow diagnosis of hyperplastic and tumorous lymph nodes by MR lymphography.

3.2.3. Charge-induced

The charges on the surface of the magnetic nanoparticles may influence their cellular interactions and the pathways/mechanisms of uptake. Positively charged iron oxide nanoparticles tend to directly bind to oppositely charged surfaces of cells, whereas negatively charged iron oxide nanoparticles may be internalized by endocytosis, particularly protein-mediated phagocytosis and diffusion. Surface modification of magnetite nanoparticles with aminosilanes (magnetic core diameter 10 nm, hydrodynamic diameter 30 nm) afford numerous positively charged surface groups under physiological conditions. Cellular uptake of these aminated-magnetite nanoparticles in glioblastoma cells was 500–2000 times higher than in normal cells [93]. Similarly, anionic nanoparticles have also found applications in cellular MR imaging, in particular, for *in-vivo* tracking of T-lymphocytes, with direct implications for cell-based anticancer therapy [94].

3.2.4. Reticuloendothelial system (RES)

Passive targeting can also occur through the inherent clearance by the RES. Comprised of bone marrow progenitors, blood monocytes, and tissue macrophages, the uptake of MNPs by these phagocytic cells provides a means of delivering contrast agents to related organs. This RES-mediated targeting was the basis for the first clinical application of MNPs in the form of Ferumoxides AMI-25 (Endorem® and Feridex IV®) for liver imaging [95]. Uptake of these MNPs by Kupffer cells of healthy hepatic parenchyma allows their differentiation from diseased tissue by the contrast-enhancement observed under MRI [96].

3.3. Active Targeting

Designing molecules that bind to targets that become up-regulated or over-expressed as normal cells become cancerous is an important strategy for both therapeutic and diagnostic drug design. One important approach to deliver drugs to any desired target involves the functionalization of the surface of nanoparticles with monoclonal antibodies or ligands to tumor-related receptors, taking advantage of the specific binding ability between an antibody and antigen, or between the ligand and its receptor. Within this approach, several differences between cancerous and normal cells can be exploited including: uncontrolled proliferation, insensitivity to negative growth regulation and antigrowth signals, angiogenesis and metastasis. It is well known that cancer cells, unlike normal cells, are rapidly proliferating. One mechanism underlying this growth is the over-expression of receptors that allows the uptake of growth factors via receptor-mediated endocytosis more efficiently than normal cells. This could be use as a "trojan horse" to deliver anticancer agents, decorating the surface of nanoparticles with antibodies or ligands that specifically bind to these receptors. The current section will review some specific examples on active targeting of MNPs with special emphasis on cancer diagnosis and imaging.

3.3.1. Antibodies

Monoclonal antibodies (mAbs) were the first targeting agents to exploit molecular recognition for delivering MNPs to their target pathological zone, and continue to be used because of their high specificity [97-99]. Recently, the development of Herceptin™, an FDA-approved mAb to the HER2/neu (erbB2) receptor, has made it a popular targeting agent for nanoparticles [100]. Specific delivery of Herceptin™-targeted DMSA-coated magnetite nanoparticles to NIH3T6.7 cells, expressing the HER2/*neu* cancer marker *in vivo*, have been demonstrated [101]. MR imaging of mice bearing xenograft tumors showed a *T2* decrease of 20% as a result of accumulation of this nanoprobe. Antibody-conjugated iron oxide nanoparticles offer a specific and sensitive tool to enhance magnetic resonance (MR) images of both local and metastatic cancer. Receptor-mediated endocytosis of a multimodal (MR/optical) imaging probe comprising of streptavidin-labeled iron oxide nanoparticles conjugated with biotinylated anti-prostate-specific membrane antigen (PSMA) antibody into LNCaP prostate cancer cells, has been demonstrated by Serda et al [102]. Nanoparticle internalization through receptor-mediated endocytosis involved the formation of clathrin-coated vesicles. Endocytosed particles were not targeted to the golgi apparatus for recycling, but instead accumulated within lysosomes. In *T1*-weighted MR images, the signal enhancement owing to the magnetic particles was greater for cells with magnetic particles bound to the cell surface than for cells that internalized the particles. However, the location of the particles (surface vs internal) did not significantly alter their effect on *T2*-weighted images. Very recently, Kou et al have developed a targeted contrast agent (SM-UMNP) by conjugating co-precipitated ULTRA MNP to a humanized SM5-1 antibody which can specifically react with human hepatocellular carcinoma (HCC) cells [103]. These SM-ULTRA MNP nanoparticles were demonstrated to be able to selectively accumulate in the tumor cells, resulting in a marked decrease of MRI *T2*-weighted signal intensity. Biodistribution studies demonstrated the efficient accumulation of SMU LTRA MNP in the ch-hep-3 tumor in nude mice. MRI studies indicated that SM-U LTRA MNP had the potential to be a promising targeted contrast agent for diagnosis of HCC. Cetuximab-conjugated fluorescent magnetic nanohybrids (CET-

FMNHs), composed of a magnetic $MnFe_2O_4$ nanocore, encapsulated in pyrene-labeled PCL-b-PMAA as a surfactant, were synthesized by a nano-emulsion method and used for the detection of human epithelial cancer via magnetic resonance (MR) and optical imaging [104]. Another interesting example of these antibody-conjugated nanomagnetic probes is provided by the recently developed epidermal growth factor receptor (EGFR)-targeted nanoparticles. This nanoprobe is synthesized by conjugating a single-chain anti-EGFR antibody (ScFvEGFR) to surface-functionalized quantum dots (QDs) or magnetic iron oxide (IO) nanoparticles, which specifically bind to and are internalized by EGFR-expressing cancer cells, thereby producing a fluorescent signal or magnetic resonance imaging (MRI) contrast into an orthotopic pancreatic cancer model [105]. Inspite of their widespread applicability, one drawback of mAbs is their large size and inherent immunogenicity, which can cause conjugated nanoparticles to diffuse poorly through biological barriers [106,107].

3.3.2. Aptamer

Another important class of affinity ligands are the aptamers, which rival antibodies in their diagnostic potential. Aptamers are oligonucleotides derived from an *in vitro* evolution process called SELEX (Systematic Evolution of Ligands by Exponential Enrichment). Aptamers are evolved to bind proteins associated with a number of diseased states. The circulation time can be considerably improved by conjugating the low molecular weight aptamers (8000-12000) to higher molecular weight vehicles. Highly specific DNA aptamers are selected by SELEX to bind with specific molecular or cellular targets. Although further optimization is necessary for application in clinical diagnostics, aptamer-conjugated magnetofluorescent nanoparticles have shown considerable promise in the specific recognition of CCRF-CEM acute leukemia cells from complex mixtures, including whole blood samples [108]. The use of aptamer-targeted nanoparticles for the collection and detection of multiple cancer cells have been further extended by Smith et al [109].

3.3.3 Peptides

3.3.3.1. Tumor-homing peptides

Targeting of MNPs to receptors overexpressed on tumor neovasculature has currently become an extensive area of research. The formation of new blood vessels, or angiogenesis, is an essential component of tumor growth and has been shown to be highly specific for neoplasia [110]. A relatively large number of angiogenesis markers, which include the $\alpha v \beta 3$ integrin, vascular endothelial growth factor (VEGF), cell surface nucleolin and heparin sulfates, have been identified as potential targets for the delivery of diagnostic and therapeutic agents [111,112]. Targeting agents, such as the Arg-Gly-Asp (RGD) peptide demonstrating high affinity for the $\alpha v \beta 3$ integrin, have been evaluated for the delivery of MNPs to a variety of neoplastic tissues including breast tumors, malignant melanomas and squamous cell carcinomas [113-115]. In a recent study by Reddy et al, a vascular homing peptide, F3, which binds to nucleolin expressed on tumor endothelium and cancer cells, was utilized to deliver a multifunctional MNP to brain tumors [116]. Simberg et al synthesized the tumor-homing peptide CREKA (Cys-Arg-Glu-Lys-Ala), and showed that when CREKA was linked on the surface of nanoparticles, they accumulated not only in the cancer site but also in the associated blood vessels, causing intravascular clotting, which suggests a potential to enhance detection strategies by optical imaging [117].

3.3.3.2. Chlorotoxin for targeting tumors of brain/neuroectodermal origin

Chlorotoxin (CTX), purified from scorpion venom, has been used to target cancers of neuroectodermal origin with success [118]. The conjugation of CTX can enhance the delivery of magnetic nanoparticles in gliomas [119]. Recently, Sun et al [2008] have reported polyethylene glycol (PEG) surface-modified magnetic nanoparticles which actively target gliomas using chlorotoxin (CTX). The presence of chlorotoxin allows the nanoparticle to target the membrane-bound matrix, metalloproteinase-2 (MMP-2) protein complex, which is up-regulated in gliomas [120].

3.3.3.3. Bombesin-targeted CLIO contrast agents for imaging pancreatic ductal adenocarcinoma

Another important target is constituted by the bombesin (BN) receptors, present on normal acinar cells of the pancreas, and have been used to image pancreatic ductal adenocarcinomas when conjugated on nanoparticles [121].

3.3.3.4. Imaging hepsin expression in prostate cancer using IPL-F conjugated, fluorescent CLIO

Hepsin, a serine protease, has been used for prostate cancer imaging. It is expressed in various lesions including precancerous, high-grade prostate intra-epithelial neoplasia, and metastatic cancers which are refractory to hormonal therapy and hence has been used as a target for improving the specificity of magnetic nanoparticles for prostate cancer imaging.

3.3.3.5. EPPT for targeting uMUC-1 over-expressing cancer cell-lines

Under-glycosylated mucin-1 antigen (uMUC-1) can act as a marker for cancers, as it is overexpressed in more than 50% of human cancers whereas normal tissues have shown high levels of glycosylation. Medarova et al have reported on a dual-modality imaging probe specifically targeting the uMUC-1. This probe consisted of three components: (i) a CLIO core for MR imaging (ii) a Cy5.5 dye for near infrared optical fluorescence imaging (NIRF), and (iii) peptides (EPPT), specifically recognizing uMUC-1, attached to the nanoparticle's dextran coat. In addition to the detection of orthotopically implanted preclinical models of adenocarcinomas, this probe could also track tumor response to chemotherapy *in vivo* in real time [122].

3.3.3.6. LHRH-conjugated ULTRA MNP for in-vivo cancer diagnosis/ imaging

Binding sites for luteinizing hormone-releasing hormone (LHRH) (now known in genome and microarray databases as GNRH1), is found on 52 % of human breast cancers [123], about 80% of human ovarian and endometrial cancers, and 86% of human prostatic carcinoma specimens. As LHRH receptors are not expressed on most normal tissues, they represent a specific target for cancer chemotherapy with antineoplastic agents linked to an LHRH vector molecule. LHRH is a decapeptide having the primary sequence of EHWSYGLRPG which can be used to target hormone-responsive breast cancers. LHRH-MNPs have been found to facilitate a 12-fold higher accumulation of the tagged nanoparticles in cancerous cells *in vitro*, and 7.5 to 11-fold higher accumulation in tumors *in vivo* as compared to the naon-targeted nanoparticles [124]. These nanoparticles also appear to be promising for the detection of metastases and disseminated cells in lymph nodes, bones and peripheral organs.

3.3.3.7. uPAR- targeted MNPs in the detection of pancreatic cancer

The urokinase plasminogen activator (uPA) system is promising for the imaging of pancreatic cancers. It comprises several components like the serine protease uPA, its receptor, uPAR, and two inhibitors, namely plasminogen activator inhibitor-1 (PAI-1) and plasminogen activator inhibitor-2 (PAI-2). Yang et al has reported a uPAR-targeted dual-modality molecular imaging nanoparticle probe, prepared by conjugating a near-infrared dye-labeled amino-terminal fragment of the receptor, binding the domain of the urokinase plasminogen activator to the surface of functionalized magnetic iron oxide nanoparticles. On systemic administration of uPAR-targeted nanoparticles, Yang et al found that they tend to accumulate within pancreatic cancers [125].

3.4. Small Molecules as Targeting Agents

Among the different strategies so far developed for receptor-mediated targeting of magnetic nanoparticles, the receptor of folic acid constitutes a useful target for tumor-specific drug delivery. Numerous attempts have been made to develop biodegradable hydrophilic magnetic nanoparticle–folate conjugate delivery systems using dextran, poly(D, L-lactic-*co*-glycolic acid) (PLGA), poly(ethylene glycol) (PEG), PLL(poly-L-Lysine) and their block copolymers [126-128]. Although these delivery systems improve biocompatibility, resist protein adsorption and increase the circulation time and internalization efficiency, the large size of these macromolecule-based systems does not facilitate intravenous delivery. So an effective design of the nanoparticle–folate-conjugate system must take into account its size and colloidal stability in the physiological environment which promises the easy perfusion of the nanoparticle system out of the blood stream to reach the target cell of interest. Kohler et al have demonstrated the sustained release of methotrexate (MTX) in breast and brain tumor cells delivered by a non-polymer based iron-oxide nano-formulation [129]. In this study, the authors covalently attached MTX to 3-aminopropyl trimethoxy-silane (APTS)-grafted magnetite nanoparticles through amidation between the γ-carboxyl groups of MTX and surface-pendant amino groups. Through the use the covalent linkage, the group demonstrated the controlled release of MTX to the cellular cytosol and the subsequent cytotoxicity to these cancer cells. Cleavage of the MTX from the MNPs was evaluated over a range of pH values and in the presence of lysozymes to mimic conditions present in the lysosomal compartments.

In our laboratory, a novel technique to synthesize highly stable folic acid (FA)-conjugated magnetite (Fe3O4) nanoparticles for targeting cancer cells using 2-Carboxyethyl phosphonic acid (CEPA) surface-coupling agent and a non-polymeric hydrophilic linker, 2, 2'-(ethylenedioxy)-bis-ethylamine (EDBE), was developed [130]. The reported iron-oxide folate nanoconjugate (Fe3O4-CEPA-FA) has been found to be non-cytotoxic, and showed high site-specific intracellular uptake against folate receptor (FR) over expressing cancer cells. Our laboratory has also developed phosphonic acid-based methodology along with conventional bioconjugation strategies to develop biocompatible, functionalized, rhodamine isothiocyanate (RITC)-tagged magnetic nanoparticles having mean diameter 10 ± 2 nm [131]. Further, folic acid was conjugated on the surface of the nanoparticles for active targeting.

4. Conjugation Agents for MNPs

A number of chemical approaches have been used for the conjugation of targeting, therapeutic and imaging-reporter molecules with nanoparticle surfaces. These can be categorized into covalent linkage strategies (direct nanoparticle conjugation, click chemistry, covalent linker chemistry) and physical interactions (electrostatic, hydrophilic/ hydrophobic, affinity interactions). The choice of chemistry is dictated, in part, by the chemical properties and functional groups found on the MNP coating and ligand to be linked. The primary goal is to bind the targeting, imaging or therapeutic moiety without compromising its functionality once attached. Functionality in such assemblies is dictated by the nature of the ligand (eg conformation of biomolecules), and the manner in which it is attached.

Bioconjugation of various chemical moieties are done for various applications of magnetic nanoparticle using the surface-free functional groups. These moieties are generally drugs or biological molecules used for delivery into *in vivo* or *in vitro* systems. There are a number of reactive ligands which are used for the chemical conjugations. Reactive ligands like anhydride, succinimidyl ester, epoxide and isothiocynate can be used for amine functional groups on the MNP for binding of dyes, proteins or antibodies. Ligands like maleimide and pyridyl disulfide can be used for sulfhydryl functional group. Amines can be used for aldehydes and active hydrogen groups, and hydrazine for aldehyde alone. Azide ligand can be used for alkyne groups. Carboxylic acid on the surfaces of nanoparticles can be treated with 1-ethyl-3-(dimethylaminopropyl) carbodiimide (EDC) and *N*-hydroxysuccinimide (NHS) or *N*-hydroxysulfosuccinimide (sulfo-NHS) and then a diamine to form amine-functionalized nanoparticles. A schematic diagram is presented in [Figure 3].

5. Applications of MNPs

5.1 Drug delivery

MNPs can be used by an external magnetic field to increase site-specific delivery of therapeutic agents [132,133]. In general, this process involves the attachment of a cytotoxic drug to a biocompatible magnetic targeted carrier (MTC), intravenous injection of these MTCs in the form of a colloidal suspension, application of a magnetic field gradient to direct the MTC to the pathological site, and release of the therapeutic agent from the MTC.

5.2 Cancer imaging

MNPs have been used as MRI contrast agents to improve the detection, diagnosis and therapeutic management of various tumors. Currently, clinical imaging of liver tumors and metastases through RES-mediated uptake of MNPs has been used to detect tumors as small as 2–3 mm [134,135]. In addition, supermagnetic nanoparticles (SMNP) have been demonstrated to successfully identify diameter of 5–10mm metastases of lymph node under MRI. Another clinical application of SMNPs under evaluation is their use in improving the delineation of brain tumor boundaries and quantifying tumor volumes [136,137].

5.3 Molecular imaging

Due to their ability to serve as molecularly targeted imaging agents, MNPs are now and will be an integral part in this biomedical field. MNPs can be used in the imaging of cell migration/trafficking [138], apoptosis detection [139] and imaging of enzyme activities [140-142].

5.4 Cardiovascular disease imaging

MNPs have been proposed as MRI contrast agents for several medical applications in cardiovascular medicine, including myocardial injury, atherosclerosis and other vascular disease [143]. The uptake of MNPs by macrophages, which have been shown to be a marker of unstable atheromatous plaques [134,144], has also been exploited to visualize these lesion-prone arterial sites. Clinical studies have demonstrated that MR imaging using SMNPs may be useful in evaluating the risk of acute ischaemic events [145,146]. Recently, 30 families of new peptides have been identified that bind to atherosclerotic lesions, through *in vivo* phage display [147].

Reference

1. Fowler, Michael. "Historical Beginnings of Theories of Electricity and Magnetism",(1997)

2. Vowles, Hugh P. "Early Evolution of Power Engineering", (University of Chicago Press), 17 no.2 (1932): 412–420

3. Li, Shu-hua. "Origine de la Boussole 11. Aimant et Boussole," Isis 45, no. 2 (1954): 175.

4. Lu, A.H. Salabas, E. L and Schüth, F Angew. Chem., Int. Ed. 46, (2007): 1222–1244

5. Sun, C. Lee, J. S. M. Zhang, M. Q. Magnetic nanoparticles in MR imaging and drug delivery. Advanced Drug Delivery Reviews 60, (2008): 1252–1265.

6. McCarthy, J. R. Weissleder, R. Multifunctional magnetic nanoparticles for targeted imaging and therapy. Advanced Drug Delivery Reviews 60, (2008): 1241–1251.

7. Lewinski, N. Colvin, V. Drezek, R. Cytotoxicity of nanoparticles. Small 4, (2008): 26–49.

8. Wang, Y.X.J. Hussain, S.M. Krestin, G.P. Superparamagnetic iron oxide contrast agents: physicochemical characteristics and applications in MR imaging. European Radiology 11, (2001): 2319–2331.

9. Lawrence, R. Development and comparison of iron dextran products, PDA. Journal of Pharmaceutical Science and Technology 52, (1998): 190–197.

10. Gupta, A.K. Gupta, M. Synthesis and surface engineering of iron oxide nanoparticles for biomedical applications. Biomaterials 26, (2005): 3995–4021.

11. Gupta, A.K. Naregalkar, R.R. Vaidya, V.D. Gupta, M. Recent advances on surface engineering of magnetic iron oxide nanoparticles and their biomedical applications. Nanomedicine 2,(2007): 23–39.

12. Laurent, S. Forge, D. Port, M. Roch, A, Robic, C. Elst, L.V. Muller, R.N. Magnetic ironoxide nanoparticles: Synthesis, stabilization, vectorization, physicochemical characterizations, and biological applications. Chemical Reviews 108, (2008): 2064–2110.

13. McCarthy, J.R. Kelly, K.A. Sun, E.Y. Weissleder, R. Targeted delivery of multifunctional magnetic nanoparticles. Nanomedicine 2, (2007): 153–167.

14. Dobson, J. Magnetic nanoparticles for drug delivery. Drug Development Research 67, (2006): 55–60.

15. Pankhurst, Q.A. Connolly, J. Jones, S.K. Dobson, J. Applications of magnetic nanoparticles in biomedicine. Journal of Physics. D. Applied Physics 36, (2003): 167–181.

16. Zhao, Z.L. Bian, Z.Y. Chen, L.X. He, X.W. Wang, Y.F. Synthesis and surfacemodifications of iron oxide magnetic nanoparticles and applications on separation and analysis. Progress in Chemistry 18,

(2006): 1288–1297.

17. Perez, J.M. Josephson, L. O'Loughlin, T. Hogemann, D. Weissleder, R. Magnetic relaxation switches capable of sensing molecular interactions. Nature Biotechnology, 20 (2002): 816–820.

18. Frullano, L. Meade, T.J. Multimodal MRI contrast agents. Journal of Biological Inorganic Chemistry 12, (2007): 939–949.

19. Corot, C. Robert, P. Idee, J.M. Port, M. Recent advances in iron oxide nanocrystal technology for medical imaging. Advanced Drug Delivery Reviews 58, (2006): 1471 1504.

20. Duran, J.D.G. Arias, J.L. Gallardo, V. Delgado, A.V. Magnetic colloids as drug vehicles. Journal of Pharmaceutical Sciences 97, (2008): 2948–2983.

21. Solanki, A. Kim, J.D. Lee, K.B. Nanotechnology for regenerative medicine: nanomaterials for stem cell imaging. Nanomedicine 3, (2008): 567–578.

22. Thiesen, B. Jordan, B. A. Clinical applications of magnetic nanoparticles for hyperthermia. International Journal of Hyperthermia 24, (2008): 467–474.

23. Bonnemain, B. Superparamagnetic agents in magnetic resonance imaging: Physicochemical characteristics and clinical applications - A review. Journal of Drug Targeting 6, (1998): 167–174.

24. Harisinghani, M.G. Barentsz, J. Hahn, P.F. Deserno, W.M. Tabatabaei, S. van de Kaa, C.H. de la Rosette, J. Weissleder, R. Noninvasive detection of clinically occult lymph-node metastases in prostate cancer. New England Journal of Medicine 348, (2003): 2491-U5.

25. Singh, A. Patel, T. Hertel, J. Bernardo, M. Kausz, A. Brenner, L. Safety of Ferumoxytol in Patients With Anemia and CKD. American Journal of Kidney Diseases 52, (2008): 907–915.

26. Lee, J.H. Huh, Y.M. Jun, Y. Seo, J. Jang, J. Song, H.T. Kim, S. Cho, Yoon, H.G. Suh, J.S. Cheon, J. Artificially engineered magnetic nanoparticles for ultra-sensitive molecular imaging. Nature Medicine 13, (2007): 95–99.

27. Goya, G.F. Grazu, V. Ibarra, M.R. Magnetic nanoparticles for cancer therapy. Current Nanoscience 4, (2008): 1–16.

28. Jurgons, R. Seliger, C. Hilpert, A. Trahms, L. Odenbach, S. Alexiou, C. Drug loaded magnetic nanoparticles for cancer therapy. Journal of Physics. Condensed Matter 18, (2006): 2893–2902.

29. Chertok, B. Moffat, B.A. David, A.E. Yu, F.Q. Bergemann, C. Ross, B.D. Yang, V.C. Iron oxide nanoparticles as a drug delivery vehicle for MRI monitored magnetic targeting of brain tumors. Biomaterials 29, (2008): 487–496.

30. Tartaj, P. Morales, M.D. Veintemillas-Verdaguer, S. Gonzalez-Carreno, T. Serna, C.J. The preparation of magnetic nanoparticles for applications in biomedicine. Journal of Physics. D. Applied Physics 36, (2003): 182–197.

31. Ferrari, M. Cancer nanotechnology: Opportunities and challenges. Nature Reviews. Cancer 5, (2005): 161–171.

32. Belting, M. Sandgren, S. Wittrup, A. Nuclear delivery of macromolecules: barriers and carriers. Advanced Drug Delivery Reviews 57, (2005): 505–527.

33. Davis, M.E. Non-viral gene delivery systems. Current Opinion in Biotechnology 13, (2002): 128–131.

34. Longmire, M. Choyke, P.L. Kobayashi, H. Clearance properties of nano-sized particles and molecules as imaging agents: considerations and caveats. Nanomedicine 3, (2008): 703–717.

35. Begley, D.J. Delivery of therapeutic agents to the central nervous system: the problems and the possibilities. Pharmacology & Therapeutics 104, (2004): 29–45.

36. Janes, K.A. Calvo, P. Alonso, M.J. Polysaccharide colloidal particles as delivery systems for macromolecules. Advanced Drug Delivery Reviews 47, (2001): 83–97.

37. Kumar, M. Muzzarelli, R.A.A. Muzzarelli, C. Sashiwa, H. Domb, A.J. Chitosan chemistry and pharmaceutical perspectives. Chemical Reviews 104, (2004): 6017–6084.

38. Kim, E.H. Ahn, Y. Lee, H.S. Biomedical applications of superparamagnetic iron oxide nanoparticles encapsulated within chitosan. Journal of Alloys and Compounds 434, (2007): 633–636.

39. Li, B.Q. Jia, D.C. Zhou, Y. Hu, Q.L. Cai, W. In situ hybridization to chitosan/magnetite nanocomposite induced by the magnetic field. Journal of Magnetism and Magnetic Materials 306, (2006): 223–227.

40. Sipos, P. Berkesi, O. Tombacz, E. St Pierre, T.G. Webb, J. Formation of spherical iron (III) oxyhydroxide nanoparticles sterically stabilized by chitosan in aqueous solutions. Journal of Inorganic Biochemistry 95, (2003): 55–63.

41. Bhattarai, S.R. Badahur, K.C.R. Aryal, S. Khil, M.S. Kim, H.Y. N-Acylated chitosan stabilized iron oxide nanoparticles as a novel nano-matrix and ceramic modification. Carbohydrate Polymers 69, (2007): 467–477.

42. Bhattarai, S.R. Kim, S.Y. Jang, K.Y. Lee, K.C. Yi, H.K. Lee, D.Y. Kim, H.Y. Hwang, P.H. Laboratory formulated magnetic nanoparticles for enhancement of viral gene expression in suspension cell line. Journal of Virological Methods 147, (2008): 213–218.

43. Kim, E.H. Lee, H.S. Kwak, B.K. Kim, B.K. Synthesis of ferrofluid with magnetic nanoparticles by sonochemical method for MRI contrast agent. Journal of Magnetism and Magnetic Materials 289, (2005): 328–330.

44. Lee, H.S. Kim, E.H. Shao, H.P. Kwak, B.K. Synthesis of SPIO-chitosan microspheres for MRI-detectable embolotherapy. Journal of Magnetism and Magnetic Materials 293, (2005): 102–105.

45. Kircheis, R. Wightman, L. Wagner, E. Design and gene delivery activity of modified polyethylenimines. Advanced Drug Delivery Reviews 53, (2001): 341–358.

46. Godbey, W.T. Wu, K.K. Mikos, A.G. Tracking the intracellular path of poly (ethylenimine)/DNA complexes for gene delivery. Proceedings of the National Academy of Sciences of the United States of America 96, (1999): 5177–5181.

47. Steitz, B. Hofmann, H. Kamau, S.W. Hassa, P.O. Hottiger, M.O. Rechenberg, B. V. Amtenbrink, M. H. Fink, A. P. Characterization of PEI-coated superparamagnetic iron oxide nanoparticles for transfection: Size distribution, colloidal properties and DNA interaction. Journal of Magnetism and Magnetic Materials 311, (2007): 300–305.

48. Chorny, M. Polyak, B. Alferiev, I.S. Walsh, K. Friedman, G. Levy, R.J. Magnetically driven plasmid DNA delivery with biodegradable polymeric nanoparticles. FASEB Journal 21, (2007): 2510–2519.

49. Park, I.K. Ng, C.P. Wang, J. Chu, B. Yuan, C. Zhang, S. Pun, S.H. Determination of nanoparticle vehicle unpackaging by MR imaging of a T-2 magnetic relaxation switch. Biomaterials 29, (2008): 724–732.

50. McBain, S.C. Yiu, H.H.P. Haj, A. E. Dobson, J. Polyethyleneimine functionalized iron oxide nanoparticles as agents for DNA delivery and transfection. Journal of Materials Chemistry 17, (2007): 2561–2565.

51. Corti, M. Lascialfari, A. Marinone, A. Masotti, A. Micotti, E. Orsini, Ortaggi, G.. Poletti, G Innocenti, C. Sangregorio, C. Magnetic and relaxometric properties of polyethylenimine-coated superparamagnetic MRI contrast agents. Journal of Magnetism and Magnetic Materials 320, (2008): 316– 319.

52. Huth, S. Lausier, J. Gersting, S.W. Rudolph, C. Plank, C. Welsch, U. Rosenecker, J. Insights into the mechanism of magnetofection using PEI-based magnetofectins for gene transfer. Journal of Gene Medicine 6, (2004): 923–936.

53. Duan, H.W. Kuang, M. Wang, X.X. Wang, Y.A. Mao, H. Nie, S.M. Reexamining the effects of particle size and surface chemistry on the magnetic properties of iron oxide nanocrystals: New insights into spin disorder and proton relaxivity. Journal of Physical Chemistry C 112,

(2008): 8127–8131.

54. Xie, J. Xu, C. Kohler, N. Hou, Y. Sun, S. Controlled PEGylation of monodisperse Fe3O4 nanoparticles for reduced non-specific uptake by macrophage cells. Advanced Materials 19, (2007): 3163–3166.

55. Kohler, N. Fryxell, G.E. Zhang, M.Q. A bifunctional poly (ethylene glycol) silane immobilized on metallic oxide-based nanoparticles for conjugation with cell targeting agents. Journal of the American Chemical Society 126, (2004): 7206–7211.

56. Kim, D.K. Zhang, Y. Kehr, J. Klason, T. Bjelke, B. Muhammed, M. Characterization and MRI study of surfactant-coated superparamagnetic nanoparticles administered into the rat brain. Journal of Magnetism and Magnetic Materials 225, (2001): 256–261.

57. Tiefenauer, L.X. Tschirky, A. Kuhne, G. Andres, R.Y. In vivo evaluation of magnetite nanoparticles for use as a tumor contrast agent in MRI. Magnetic Resonance Imaging 14, (1996): 391–402.

58. Illum, L. Church, A.E. Butterworth, M.D. Arien, A. Whetstone, J. Davis, S.S. Development of systems for targeting the regional lymph nodes for diagnostic imaging: In vivo behaviour of colloidal PEG-coated magnetite nanospheres in the rat following interstitial administration. Pharmaceutical Research 18, (2001): 640–645.

59. Papisov, M.I. Bogdanov, A. Schaffer, B. Nossiff, N. Shen, T. Weissleder, R. Brady, T.J., colloidal magnetic-resonance contrast agents - effect of particle surface on biodistribution. Journal of Magnetism and Magnetic Materials 122, (1993): 383–386.

60. Sun, C. Fang, C. Stephen, Z. Veiseh, O. Hansen, S. Lee, D. Ellenbogen, R.G. Olson, J. Zhang, M.Q. Tumor-targeted drug delivery and MRI contrast enhancement by chlorotoxin- conjugated iron oxide nanoparticles. Nanomedicine 3, (2008): 495–505.

61. Sun, C. Veiseh, O. Gunn, J. Fang, C. Hansen, S. Lee, D. Sze, R. Ellenbogen, R.G. Olson, J. Zhang, M. In vivo MRI detection of gliomas by chlorotoxin-conjugated superparamagnetic nanoprobes. Small 4, (2008): 372–379.

62. Chen, X. Zhang, W. Laird, J. Hazen, S.L. Salomon, R.G. Polyunsaturated phospholipids promote the oxidation and fragmentation of gamma-hydroxyalkenals: formation and reactions of oxidatively truncated ether phospholipids. Journal of Lipid Research 49, (2008): 832–846.

63. Lutz, J.F. Stiller, S. Hoth, A. Kaufner, L. Pison, U. Cartier, R. One-pot synthesis of PEGylated ultrasmall iron-oxide nanoparticles and their invivo evaluationasmagnetic resonance imaging contrast agents. Biomacromolecules 7, (2006): 3132–3138.

64. Veiseh, O. Gunn, J. Kievit, F. Sun, C. Fang, C. Lee, Zhang, J. Inhibition of tumor cell invasion with chlorotoxin-bound superparamagnetic nanoparticles. Small 5, (2009): 256–264.

65. Josephson, L. Tung, C.H. Moore, A. Weissleder, R. High-efficiency intracellular magnetic labeling with novel superparamagnetic-tat peptide conjugates. Bioconjugate Chemistry 10, (1999): 186–191.

66. Wunderbaldinger, P. Josephson, L. Weissleder, R. Crosslinked iron oxides (CLIO): A new platform for the development of targeted MR contrast agents. Acad Radiol 9, (2002): 304–306.

67. Bulte, J.W.M. Ma, L.D. Magin, R.L. Kamman, R.L. Hulstaert, C.E. Go, K.G. Deleij, L. selective MR imaging of labeled human peripheral-blood mononuclear-cells by liposome mediated incorporation of dextran-magnetite particles. Magnetic Resonance in Medicine 29, (1993): 32–37.

68. Martina, M.S. Fortin, J.P. Menager, C. Clement, O. Barratt, G. Madelmont, C. G. Gazeau, F. Cabuil, V. Lesieur, S. Generation of superparamagnetic liposomes revealed as highly efficient MRI contrast agents for in vivo imaging. Journal of the American Chemical Society 127, (2005): 10676–10685.

69. Bulte, J.W.M. Cuyper, M. D. Despres, D. Frank, J.A. Short- vs. long-circulating magnetoliposomes as bone marrow-seeking MR contrast agents. Journal of Magnetic Resonance Imaging 9, (1999): 329–335.

70. Decuyper, M. Joniau, M. magnetoliposomes - formation and structural characterization. European Biophysics Journal with Biophysics Letters 15, (1988): 311–319.

71. Yang, J. Lee, T.I. Lee, J. Lim, E.K. Hyung, W. Lee, C.H. Song, Y.J. Suh, J.S. Yoon, H.G. Huh, Y.M. Haam, S. Synthesis of ultrasensitive magnetic resonance contrast agents for cancer imaging using PEG-fatty acid. Chemistry of Materials 19, (2007): 3870–3876.

72. Mulder, W.J.M. Strijkers, G.J. Tilborg, G.A.F.V. Griffioen, A.W. Nicolay, K. Lipidbased nanoparticles for contrast-enhanced MRI and molecular imaging. NMR in Biomedicine 19, (2006): 142–164.

74. Zhang, Y. Kohler, N. Zhang, M.Q. Surface modification of superparamagnetic magnetite nanoparticles and their intracellular uptake. Biomaterials 23, (2002): 1553–1561.

75. Sinha, R. Kim, G.J. Nie, S.M. Shin, D.M. Nanotechnology in cancer therapeutics: bioconjugated nanoparticles for drug delivery. Molecular Cancer Therapeutics 5, (2006): 1909–1917.

76. Kohler, N. Sun, C. Wang, J. Zhang, M.Q. Methotrexate-modified superparamagnetic nanoparticles and their intracellular uptake into human cancer cells. Langmuir 21, (2005): 8858–8864.

77. Sun, C. Sze, R. Zhang, M.Q. Folic acid-PEG conjugated superparamagnetic nanoparticles for targeted cellular uptake and detection by MRI. Journal of Biomedical Materials Research Part A 78, (2006): 550–557.

78. Veiseh, O. Sun, C. Gunn, J. Kohler, N. Gabikian, P. Lee, D. Bhattarai, N. Ellenbogen, R. Sze, R. Hallahan, A. Olson, J. Zhang, M.Q. Optical and MRI multifunctional nanoprobe for targeting gliomas. Nano Letters 5, (2005): 1003–1008.

79. Montet, X. Abou, K M. Reynolds, F. Weissleder, R. Josephson, L. Nanoparticle imaging of integrins on tumor cells. Neoplasia 8, (2006): 214–222.

80. Montet, X. Weissleder, R. Josephson, L. Imaging pancreatic cancer with a peptidenanoparticle conjugate targeted to normal pancreas. Bioconjugate Chemistry 17, (2006): 905–911.

81. Boutry, S. Laurent, S. Elst, L.V. Muller, R.N. Specific E-selectin targeting with a superparamagnetic MRI contrast agent. Contrast Media & Molecular Imaging 1, (2006): 15–22.

82. Gunn, J. Wallen, H. Veiseh, O. Sun, C. Fang, C. Cao, J.H. Yee, C. Zhang, M.Q. A multimodal targeting nanoparticle for selectively labeling T cells. Small 4, (2008): 712–715.

83. Artemov, D. Mori, N. Okollie, B. Bhujwalla, Z.M. MR molecular imaging of the Her- 2/neu receptor in breast cancer cells using targeted iron oxide nanoparticles. Magnetic Resonance in Medicine 49, (2003): 403–408.

84. Hu, F.Q. Wei, L. Zhou, Z. Ran, Y.L. Li, Z. Gao, M.Y. Preparation of biocompatible magnetite nanocrystals for in vivo magnetic resonance detection of cancer. Advanced Materials 18, (2006): 2553–2556.

85. Huh, Y.M. Jun, Y.W. Song, H.T. Kim, S. Choi, J.S. Lee, J.H. Yoon, S. Kim, K.S. Shin, J.S. Suh, J.S. Cheon, J. In vivo magnetic resonance detection of cancer by using multifunctional magnetic nanocrystals. Journal of the American Chemical Society 127, (2005): 12387–12391.

86. Schafer, R. Wiskirchen, J. Guo, K. Neumann, B. Kehlbach, R. Pintaske, J. Voth, V. Walker, T. Scheule, A.M. Greiner, T.O. Klein, U. H. Claussen, C.D. Northoff, H. Ziemer, G. Wendel, H.P. Aptamer-based isolation and subsequent imaging of mesenchymal stem cells in ischemic myocard by magnetic resonance imaging, Rofo.179, (2007): 1009–1015.

87. Yigit, M.V. Mazumdar, D. Kim, H.K. Lee, J.H. Dintsov, B. Lu, Y. Smart "Turn-on" magnetic resonance contrast agents based on aptamer-functionalized superparamagnetic iron oxide nanoparticles. ChemBioChem 8, (2007): 1675–1678.

88. Yigit, M.V. Mazumdar, D. Lu, Y. MRI detection of thrombin with aptamer functionalized superparamagnetic iron oxide nanoparticles. Bioconjugate Chemistry 19, (2008): 412–417.

89. Lubbe, A.S. Bergemann, C. Riess, H. Schriever, F. Reichardt, P. Possinger, K. Matthias, M. Dorken, B. Herrmann, F. Gurtler, R. Hohenberger, P. Haas, N. Sohr, R. Sander, B. Lemke, A.J. Ohlendorf, D. Huhnt, W. Huhn, D. Clinical experiences with magnetic drug targeting: a phase I study with 4'-epidoxorubicin in 14 patients with advanced solid tumors. Cancer Research 56, (1996): 4686–4693.

90. Alexiou, C. Arnold, W. Klein, R.J. Parak, F.G. Hulin, P. Bergemann, C. Erhardt, W. Wagenpfeil, S. Lubbe, A.S. Locoregional cancer treatment with magnetic drug targeting. Cancer Research 60, (2000): 6641–6648.

91. Grief, A.D. Richardson, G. Mathematical modelling of magnetically targeted drug delivery. Journal of Magnetism and Magnetic Materials 293, (2005): 455–463.

92. Maeda, H. The enhanced permeability and retention (EPR) effect in tumor vasculature: the key role of tumor-selective macromolecular drug targeting. Advances in Enzyme Regulation 41, (2001): 189–207.

93. Dupin, L. New generation of genetically-modified organisms. Biofutur 39, (2003): 8

94. Smirnov, P. Cellular magnetic resonance imaging using superparamagnetic anionic iron oxide nanoparticles: applications to in vivo trafficking of lymphocytes and cell-based anticancer therapy Methods. Mol Biol. 512, (2009) 333-53.

95. Clement, O. Siauve, N. Lewin, M. Kerviler, E.D. Cuenod, C.A. Frija, G. Contrast agents in magnetic resonance imaging of the liver: present and future. Biomedicine and Pharmacotherapy 52, (1998): 51–58.

96. Stark, D.D. Weissleder, R. Elizondo, G. Hahn, P.F. Saini, S. Todd, L.E. Wittenberg, J. Ferrucci, J.T. Superparamagnetic iron-oxide-clinical-application as a contrast agent for MR imaging of the liver. Radiology 168, (1988): 297–301.

97. Cerdan, S. Lotscher, H.R. Kunnecke, B. Seelig, J. Monoclonal antibody-coated magnetite particles as contrast agents in magnetic resonance imaging of tumors. Magnetic Resonance in Medicine 12, (1989): 151–163.

98. Weissleder, R. Lee, A.S. Fischman, A.J. Reimer, P. Shen, T. Wilkinson, R. Callahan, R.J. Brady, T.J. Polyclonal human immunoglobulin G labeled with polymeric iron oxide: antibody MR imaging. Radiology 181, (1991): 245–249.

99. Bulte, J.W. M. Hoekstra, Y. Kamman, R.L. Magin, R.L. Webb, A.G. Briggs, R.W. Go, K.G. Hulstaert, C.E. Miltenyi, S. Specific MR imaging of human lymphocytes by monoclonal antibody-guided dextran-magnetite particles. Magnetic Resonance in Medicine 25, (1992): 148–157.

100. Artemov, D. Mori, N. Ravi, R. Bhujwalla, Z.M. Magnetic resonance molecular imaging of the HER-2/neu receptor. Cancer Research 63, (2003): 2723–2727.

101. Huh, Y.M. Jun, Y.W. Song, H.T. Kim, S. Choi, J.S. Lee, J.H. Yoon, S. Kim, K.S. Shin, J.S. Suh, J.S. Cheon, J. In vivo magnetic resonance detection of cancer by using multifunctional magnetic nanocrystals. Journal of the American Chemical Society 127, (2005): 12387–12391.

102. Serda, R.E. Adolphi, N.L. Bisoffi, M. Targeting and cellular trafficking of magnetic nanoparticles for prostate cancer imaging. Mol Imaging 6, (2007): 277–88.

103. Kou, G. Wang, S. Cheng, C. Gao, J. Li, B. Wang, H. Qian, W. Hou, S. Zhang, D. Dai, J. Gu, H. Guo, Y. Development of SM5-1-conjugated ultrasmall superparamagnetic iron oxide nanoparticles for hepatoma detection. Biochem Biophys Res Commun. 374, (2008): 192-197.

104. Yang, J. Lim, E.K. Lee, H.J. Park, J. Lee, S.C. Lee, K. Yoon, H.G. Suh, J.S. Huh, Y.M. Haam, S. Fluorescent magnetic nanohybrids as multimodal imaging agents for human epithelial cancer detection. Biomaterials. 29, no. 16 (2008): 2548-55.

105. Yang, L. Mao, H. Wang, Y.A. Cao, Z. Peng, X. Wang, X. Duan, H. Ni, C. Yuan, Q. Adams, G. Smith, M.Q. Wood, W.C. Gao, X. Nie, S. Single chain epidermal growth factor receptor antibody conjugated nanoparticles for in vivo tumor targeting and imaging. Small 5, no.2 (2009): 235-43.

106. Jain, R.K. Transport of molecules in the tumor interstitium — a review. Cancer Research 47, (1987): 3039–3051.

107. Foon, K.A. Biological response modifiers—the new immunotherapy. Cancer Research 49, (1989): 1621–1639.

108. Herr, J.K. Smith, J.E. Medley, C.D. Shangguan, D. Tan, W. Aptamer-conjugated nanoparticles for selective collection and detection of cancer cells. Anal Chem. 78, no. 9 (2006): 2918-24.

109. Smith, J.E. Medley, C.D. Tang, Z. Shangguan, D. Lofton, C. Tan, W. Aptamer-conjugated nanoparticles for the collection and detection of multiple cancer cells. Anal Chem. 79, no. 8 (2007): 3075-82.

110. Folkman, J. Tumor angiogenesis: therapeutic implications. New England Journal of Medicine 285, (1971): 1182–1186.

111. Ruoslahti, E. Specialization of tumour vasculature. Nature Reviews Cancer 2, (2002): 83–90.

112. Neri, D. Bicknell, R. Tumour vascular targeting, Nature Reviews. Cancer 5, 436–446.

113. Sunderland, C.J. Steiert, M. Talmadge, J.E. Derfus, A.M. Barry, S.E. Targeted nanoparticles for detecting and treating cancer. Drug Development Research 67, (2006): 70–93.

114. Montet, X. Montet, A.K. Reynolds, F. Weissleder, R. Josephson, L. Nanoparticle imaging of integrins on tumor cells. Neoplasia 8, (2006): 214–222.

115. Zhang, C.F. Jugold, M. Woenne, E.C. Lammers, T. Morgenstern, B. Mueller, M.M. Zentgraf, H. Bock, M. Eisenhut, M. Semmler, W. Kiessling, F. Specific targeting of tumor angiogenesis by RGDconjugated ultrasmall superparamagnetic iron oxide particles using a clinical 1.5-T magnetic resonance scanner. Cancer Research 67, (2007): 1555–1562.

116. Reddy, G.R. Bhojani, M.S. McConville, P. Moody. J. Moffat, B.A. Hall, D.E. Kim, G. Koo, Y.E.L. Woolliscroft, M.J. Sugai, J.V. Johnson. T.D. Philbert, M.A. Kopelman, R. Rehemtulla, A. Ross, B.D. Vascular targeted nanoparticles for imaging and treatment of brain tumors. Clinical Cancer Research 12, (2006): 6677–6686.

117. Simberg, D. Duza, T. Park, J.H. et al. Biomimetic amplification of nanoparticle homing to tumors. Proc Natl Acad Sci U S A 104, (2007): 932–6.

118. Veiseh, M. Gabikian, P. Bahrami, S.B. Veiseh, O. Zhang, M. Hackman, R.C. Ravanpay, A.C. Stroud, M.R. Kusuma, Y. Hansen, S.J. Kwok, D. Munoz, N.M. Sze, R.W. Grady, W.M. Greenberg, N.M. Ellenbogen, R.G. Olson, J.M. Tumor paint: a chlorotoxin: Cy5.5 bioconjugate for intraoperative visualization of cancer foci. Cancer Research 67, (2007): 6882–6888.

119. Deshane, J. Garner, C.C. Sontheimer, H. Chlorotoxin inhibits glioma cell invasion via matrix metalloproteinase-2. Journal of Biological Chemistry 278, (2003): 4135–4144.

120. Lyons, S.A. Neal, J. O. Sontheimer, H. Chlorotoxin, a scorpion-derived peptide, specifically binds to gliomas and tumors of neuroectodermal origin. Glia 39, (2002): 162–173.

121. Medarova, Z. Pham, W. Kim, Y. Dai, G. Moore, A. In vivo imaging of tumor response to therapy using a dual-modality imaging strategy. Int J Cancer 11, (2006): 2796-802.

122. Chatzistamou, L. Schally, A.V. Nagy, A. et al. Effective treatment of metastatic MDA-MB-435 human estrogen-independent breast carcinomas with a targeted cytotoxic analogue of luteinizing hormone-releasing hormone AN-207. Clin Cancer Res 6, (2000): 4158–65.

123. Leuschner, C. Kumar, C.S. Hansel, W. Soboyejo, W. Zhou, J. Hormes, J. LHRH-conjugated magnetic iron oxide nanoparticles for detection of breast cancer metastases. Breast Cancer Res Treat. 99, no.2 (2006): 163-76.

124. Yang, J. Lim, E.K. Lee, H.J. Park, J. Lee, S.C. Lee, K. Yoon, H.G. Suh, J.S. Huh, Y.M. Haam, S. Fluorescent magnetic nanohybrids as multimodal imaging agents for human epithelial cancer detection. Biomaterials 29, no.16 (2008): 2548-55.

125. Choi, H. Choi, S.R. Zhou, R. Kung, H.F. Chen, I.W. Iron oxide nanoparticles as magnetic resonance contrast agent for tumor imaging via folate receptor-targeted delivery. Acad. Radiol. 11, (2004): 996.

126. Kim, S.H. Jeong, J.H. Joe, C.O. Park, T.G. Folate receptor mediated intracellular protein delivery using PLL–PEG–FOL conjugate. J. Control.Release 103, (2005): 625.

127. Sonvico, F. Mornet, S. Vasseur, S. Dubernet, C. Jaillard, D. Degrouard, J. Hoebeke, J. Duguet, E. Colombo, P. Couvreur, P. Folate-Conjugated Iron Oxide Nanoparticles for Solid Tumor Targeting as Potential Specific Magnetic Hyperthermia Mediators: Synthesis, Physicochemical Characterization, and in Vitro Experiments. Bioconjug. Chem. 16, (2005): 1181.

128. Kohler, N. Sun, C. Wang, J. Zhang, M.Q. Methotrexate-modified superparamagnetic nanoparticles and their intracellular uptake into human cancer cells. Langmuir 21, (2005): 8858–8864.

129. Mohapatra, S. Mallick, S.K. Maiti, T.K. Ghosh, S.K. Pramanik, P. Synthesis of highly stable folic acid conjugated magnetite nanoparticles for targeting cancer cells. Nanotechnology 18, (2007): 385102 Molday RS. US patent 4, 452,773 (1984): 17-18.

130. Das, M. Mishra, D. Maiti, T.K. Basak, A. Pramanik, P. Bio-functionalization of magnetite nanoparticles using an aminophosphonic acid coupling agent: new, ultradispersed, iron-oxide folate nanoconjugates for cancer-specific targeting. Nanotechnology 19, (2008): 415101.

131. Pankhurst, Q.A. Connolly, J. Jones, S.K. Dobson, J. Applications of magnetic nanoparticles in biomedicine, Journal of Physics. D. Applied Physics 36, (2003): 167–181.

132. Dobson, J. Magnetic nanoparticles for drug delivery. Drug Development Research 67, (2006): 55–60.

133. Corot, C. Robert, P. Idee, J.M. Port, M. Recent advances in iron oxide nanocrystal technology for medical imaging. Advanced Drug Delivery Reviews 58, (2006): 1471–1504.

134. Semelka, R.C. Helmberger, T.K. Contrast agents for MR imaging of the liver. Radiology 218, (2001): 27–38.

135. Enochs, W.S. Harsh, G. Hochberg, F. Weissleder, R. Improved delineation of human brain tumors on MR images using a long-circulating, superparamagnetic iron oxide agent. Journal of Magnetic Resonance Imaging 9, (1999): 228–232.

136. Neuwelt, E.A. Varallyay, P. Bago, A.G. Muldoon, L.L. Nesbit, G. Nixon, R. Imaging of iron oxide nanoparticles by MR and light microscopy in patients with malignant brain tumours. Neuropathology and Applied Neurobiology 30, (2004): 456–471.

137. Bulte, J.W. Kraitchman, D.L. Iron oxide MR contrast agents for molecular and cellular imaging. NMR in Biomedicine 17, (2004): 484–499.

138. Schellenberger, E.A. Sosnovik, D. Weissleder, R. Josephson, L. Magneto/optical annexin V, a multimodal protein. Bioconjugate Chemistry 15, (2004): 1062–1067.

139. Tung, C.H. Mahmood, U. Bredow, S. Weissleder, R. In vivo imaging of proteolytic enzyme activity using a novel molecular reporter. Cancer Research 60, (2000): 4953–4958.

140. Weissleder, R. Tung, C.H. Mahmood, U. Bogdanov A. Jr. In vivo imaging of tumors with protease-activated near-infrared fluorescent probes. Nature Biotechnology 17, (1999): 375–378.

141. Sosnovik, D.E. Weissleder, R. Emerging concepts in molecular MRI. Current Opinion in Biotechnology 18, (2007): 4–10.

142. Sosnovik, D.E. Nahrendorf, M. Weissleder, R. Molecular magnetic resonance imaging in cardiovascular medicine. Circulation 115, (2007): 2076–2086.

143. Wickline, S.A. Neubauer, A.M. Winter, P.M. Caruthers, S.D. Lanza, G.M. Molecular imaging and therapy of atherosclerosis with targeted nanoparticles. Journal of Magnetic Resonance Imaging 25, (2007): 667–680.

144. Corot, C. Petry, K.G. Trivedi, R. Saleh, A. Jonkmanns, C. Bas, J.F.L. Blezer, E. Rausch, M. Brochet, B. Gareau, P.F. Baleriaux, D. Gaillard, S. Dousset, V. Macrophage imaging in central nervous system and in carotid atherosclerotic plaque using ultrasmall superparamagnetic iron oxide inmagnetic resonance imaging. Investigative Radiology 39, (2004): 619–625.

145. Kooi, M.E. Cappendijk, V.C. Cleutjens, K. Kessels, A.G.H. Kitslaar, P. Borgers, M. Frederik, P.M. Daemen, M. Engelshoven, J.M.A.V. Accumulation of ultrasmall superparamagnetic particles of iron oxide in human atherosclerotic plaques can be detected by in vivo magnetic resonance imaging. Circulation 107, (2003): 2453–2458.

146. Trivedi, R.A. U-King-Im, J.M. Graves, M.J. Cross, J.J. Horsley, J. Goddard, M.J. Skepper, J.N. Quartey, G. Warburton, E. Joubert, I. Wang, L.Q. Kirkpatrick, P.J. Brown, J. Gillard, J.H. In vivo detection of macrophages in human carotid atheroma — temporal dependence of ultrasmall superparamagnetic particles of iron oxideenhanced MRI. Stroke 35, (2004): 1631–1635.

147. Kelly, K.A. Nahrendorf, M. Yu, A.M. Reynolds, F. Weissleder, R. In vivo phage display selection yields atherosclerotic plaque targeted peptides for imaging. Molecular Imaging and Biology 8, (2006): 201–207.

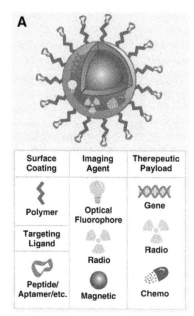

Figure 1. Graphical abstract of a magnetic nano core with surface coating and targeting agents along with delivery agents (Resource: Pramanik, P. unpublished data).

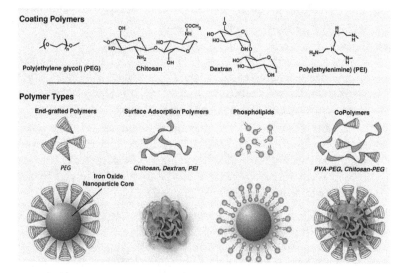

Figure 2. polymer coating of magnetic nanoparticles
(Resource: Pramanik, P. unpublished data).

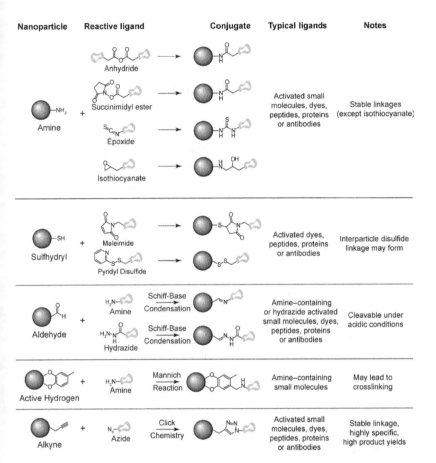

Figure 3. Ligands conjugated with magnetic nanopaticles for specific targeting and activation of biomolecules (Resource: Pramanik, P. unpublished data)

Chapter 6

BIOLOGICAL IMPLICATIONS OF POLYMER NANOCOMPOSITES

Abhijit Bandyopadhyay

Since 1959, after Dr Richard Feynman's memorable lecture, "There is Plenty of Space at the Bottom," at the American Physical Society, the scientific world has seen a boom in the area of nanoscience and nanotechnology. In the last two decades, the world of polymer science has also experienced the evolution of a new class of composites known as nanocomposites. Polymer nanocomposites are microscopically heterophase, containing embedded isotropic/anisotropic particulate entities of nanometric size within soft polymer matrices. The presence of these nanosized entities offers unmatched physico-mechanicals, chemical and rheological properties which not only excel in ordinary applications but also have shown great promise for many advanced applications in near future. Normally, very low concentration of nanofillers, sometimes 10-15 times less than conventional-sized fillers, are required to achieve success, and this has been the key to outplay the high cost factor of these materials or their precursors for popularizing/industrializing nanotechnology. The reduced size of nano-additives practically leads to high available surface for polymer adsorption. There has been a long debate on size factor of nanomaterials – generally nanodimension is defined within 1-100 nm size range – but there are several references which claim nanocomposites with size ranging high above 100 nm.

Off late, polymers have been considered an important biomaterial due to their excellent biocompatibility. All natural (including microbial polymers and natural gums) and many synthetic polymers show compatibility with living organisms, as they support cellular activity despite the fact that some of these synthetic polymers are non-biodegradable. For example, poly (vinyl alcohol) (not all grades), poly (acrylic acid), poly (hydroxyethyl methacrylate), silicone rubber etc are excellent biocompatible polymers which are not biodegradable, where biodegradability is defined as the microbial chain scission of polymers leading to mineralization. Polymers of natural origin are biodegradable. Biocompatibility is the likeliness between polymer and biological body. Few examples of biopolymers synthesized from renewable sources are poly (lactic acid), poly (hydroxyalkanoates) (a family of polymers in which poly (hydroxybutyrate) and poly (hydroxyvalerate) and their copolymers are important members), cellulose and its derivatives, starch and thermoplastic starch, chitin and chitosan, gelatin and natural rubber. Important biopolymers derived from petroleum or non-renewable sources are poly (vinyl alcohol), poly (acrylic acid), poly (N-vinyl pyrolidone), poly (acrylamide), poly (butylene succinate), aliphatic polyesters and poly (ϵ-caprolactone). Biopolymers derived from green resources usually have low mechanical strength and limited processibility, which can be remedied using nanomaterials. Synthetic biopolymers generally are devoid of such limitations.

As already stated, nanomaterials could be isotropic or anisotropic in nature. Examples of isotropic nanoparticles are metals and metal oxide nanoparticles which are spherical in dimension. A schematic of such nanoparticles is shown in Figure 1.

Isotropic nanoparticles were the first to be explored, and are considered as first generation nanoparticles. Anisotropic nanoparticles have one out of their three dimensions within nanometer scale (1-100 nm). Layered silicates and hydroxides, layered graphene, carbon nanotube, nanofibre, nanowhiskers etc. are such type of nanoparticles with their widths falling within 100 nm, but not their length. This class of nanoparticles came into picture after metals/metal oxide nanoparticles, and is regarded as second generation nanoparticles. Figure 2a shows a schematic of layered nanoparticles found in clay or graphene. The intermediate space between two subsequent layers is called gallery gap. Clay is composed of layered silicates and its gallery is filled with several adsorbed cations like sodium, potassium, magnesium, calcium etc. as charge neutralizers of silicate ions. Graphene is the layered unit of graphite, and is an assembly of π-bonded carbon atoms. Once these are folded, we get a carbon nanotube. If a single layer is folded, the resultant structure is called a single-walled carbon nanotube. For multiple layers of folding, a multi-walled carbon nanotube is obtained. Figure 2b shows this as ready reference.

Very recently, organic clusters of nanodimensions were synthesized, forming what is considered the latest or third generation nanoparticles.

Generally, nanoclusters have irregular shapes as shown in Figure 3. Tailor-made polyhedral oligomeric silsesquioxane (POSS) is an example of such type of nanomaterials.

Selection of nanoparticles for biomedical/biological uses is extremely important with regard to nanotoxicity. Nanotoxicity is an important but under-discussed topic in polymer nanotechnology, especially in the area of biomedical application. For example, carbon nanotube has been reported as a bio-toxin, so it is unwise to select carbon nanotube for such applications. Instead its functionalized form, which is less toxic, can be judiciously used. With this short backdrop, the present chapter is designed to describe general synthesis routes to polymer nanocomposites and their characterization. As already stated, there is a vast amount of literature available since from the past two decades. This chapter will only focus on to biodegradable and biocompatible polymer nanocomposites having biological implications.

Routes to Polymer-Nanocomposites Synthesis

There are three general routes to synthesize polymer nanocomposites, namely the Solution, Melt and In-situ methods [2]. Solution technique involves ultrasonic intermixing of a solvated polymer with dispersed nanoparticles [3]. Though this is an uncomplicated way of synthesizing nanocomposites, it is frequently opposed due to the ill-effects of different organic solvents. For water soluble polymers and aqueous nanoparticle sols, this is an ideal way to synthesize its nanocomposites. For example, Bhunia et al [4] have recently reported formation of high hydrolyzed-grade poly (vinyl alcohol)-nanosilica hybrids using sodium lauryl sulphate as a silica anti-coagulant. They have ultrasonically mixed aqueous poly (vinyl alcohol) with different percentages of aqueous nanosilica sol in the presence of surfactant to obtain optically transparent hybrid nanocomposites. Oliveira et al have reported the formation of weak polymer gel after intermixing non-ionic polysaccharide galactomannan with nanosilica sol [5]. Daniel de Silva et al have synthesized k-Carrageenan/nanosilica hydrogels via intermixing, and have studied its rheological behavior [6]. Guňko et al have reported starch-silica hybrid nanocomposites where silica has been treated with aminoproylmethyl silane [7]. Intimate polymer-nanoparticle adhesion is the key, and it is more difficult to achieve with anisotropic, layered nanoparticles such as silicates and graphene due to their stacked morphology. De-structuring into single/bi/tri etc. layered structure is called exfoliation (Figure 4), and its efficiency is assigned as the inverse of the average of the thickness of individual stacks dispersed within the polymer matrix. The higher the thickness of these stacks, the more intercalated the structures are. More fractional increase in exfoliated structures in a composite raises physico-mechanicals to a much higher level [8].

Use of surfactants or pretreatment of the nanomaterials itself helps in preventing large-scale aggravation of nanomaterials in solution synthesis route. Pretreatment of nanoclay is basically an ion exchange reaction carried out to replace basic cations like sodium or magnesium with organically modified ammonium or phosphonium ions so that the long organic chains can widen the clay galleries [9]. This is schematically described in Figure 5. Widening of the gallery gap is determined from wide-angle X-ray diffraction analysis. Graphite is often oxidized with strong agents like dichromate-acid or permanganate into graphite oxide [10] which has a layered nanostructure. Here, gallery widening occurs via repulsive interaction between oxygenated groups and the constituent layers. Recently, Gómez-Avilês et al synthesized a new grapheme-like nanofiller by thermally charring sucrose and sepiolite (clay) mix [11]. It is difficult to work with single-walled a carbon nanotube due to its very high surface energy, which leads to poor dispersion; the multi-walled variety is therefore often preferred. It has an inert surface and needs to be functionalized with strong acid to make it interactive. Giri et al have reported one such exercise on a carbon nanotube in which they proved the appearance of several oxygenated groups like carboxyl, hydroxyl, carbonyl, ether etc. on the nanotube surface upon mixed acid digestion [12]. Functionalization ultimately helps in better dispersion within a semi-natural polymer in later stages of processing [12]. Several instances of the formation of clay-biopolymer nanocomposites in solution have been reported. Ogata et al have synthesized PLA-clay nanocomposites from hot chloroform using different clays [13]. Chen et al reported the formation of poly (hydroxybutyrate-co-hydroxyvalerate) – organically modified clay nanocomposites via solution intercalation – and also studied the influence of clay on the crystallization kinetics of the copolymer [14]. Similarly, thermoplastic starch-modified clay has been formed via solution cast method [15]. Zheng et al have prepared gelatin-clay hybrid nanocomposites from aqueous medium [16]. Lim et al have synthesized biodegradable aliphatic polyester-based clay hybrid nanocomposites from solution using chloroform as a coagent [17]. Later, YuanQiao reported the formation of highly transparent gelatin-clay hybrid nanocomposites from aqueous phase with mechanical properties at least three times higher than those of neat, unmodified gelatin [18]. Chen et al have reported gelatin-hydroxyapatite nanorod composites formation in aqueous phase [19]. Previously, Chang et al had reported similar nanocomposite formation from aqueous medium [20]. Solution intercalation process has also been applied to synthesize high hydrolyzed grade poly (vinyl alcohol)-modified clay hybrid nanocomposites [21-23]. Recently, Yang-Su et al synthesized chitosan-clay hybrid nanocomposites via ion exchange reaction from aqueous phase [24]. Hibino reported lactate and carbonate-bound layered double hydroxide-filled agarose nanocomposites synthesized from hot aqueous solution [25]. Recently, Chiang and Wu prepared organically modified double layered hydroxide-poly (lactic acid) nanocomposites in solution [26]. Giri et al have synthesized carboxymethyl guar gum-functionalized multi-walled carbon nanotube hybrid hydogels from aqueous medium for biomedical uses [12]. In most of these cases, the matrix polymer does not possess great melt-processing character.

Melt preparation method is industrially most appreciated, but is associated with lower chances of forming good polymer-nanofiller contacts due to process limitations. Melt mixing is shear-controlled and is carried out in visco-fluidic state of polymers with limited contact time and restricted segmental movement, in contrast to the solution process where low viscosity and higher contact time ensures efficient diffusion of polymer chain segments inside intra-layers of stacked silicates/hydroxides and graphenes. More often than not the melt mixing processes are accomplished in twin screw mixers to offer high shear and chronological thick-thin melt profile. This, in actual fact, is to compensate for the low contact time limitations of the dry blending process. Additionally, compatibilizers are frequently used to ensure exfoliated morphology. Unless compatibilized, melt mixing often results in a more intercalated type of nanocomposites. There are many instances of adopting such procedures for synthesizing biopolymer-clay nanocomposites. Sinha Ray et al [27-30] and Pluta et al [31] have synthesized poly (lactic acid)-clay hybrid nanocomposites in melt. Sinha Ray et al have used the twin screw extruder, and the formation of nanocomposites has been accomplished at 190^0 C using poly (caprolactone) as compatibilizer. In a very recent reference, Fukushima et al demonstrated the formation of poly (lactic acid) and poly (caprolactone)-based fumed silica hybrid nanocomposites in melt [32]. The biodegradation properties have been investigated. Surprisingly, nanosilica protected poly (caprolactone) from degradation but accelerated it in poly (lactic acid), under identical conditions [32]. Nanocomposite from cellulose whiskers and poly (lactic acid) in melt has been reported by Oksman et al [33]. They first dispersed nanowhiskers in solvent, which had been added to poly (lactic acid) melt and extruded to achieve uniform dispersion inside the matrix. Maity et al have produced poly (hydroxybutyrate) – organically modified clay nanocomposites – in a twin screw extruder under visoc-fluidic state of the polymer (180^0 C) [34]. They have also investigated the physico-mechanicals and biodegradability of the resultant nanocomposites. Thermoplastic starch is more flexible and process-friendly than ordinary starch due to plasticization, and so its nanocomposites with modified clay can be prepared from melt. De Carvalho et al [35] and Park et al [36] both reported melt-stage formation of thermoplastic starch-based clay hybrid nanocomposites. The effects of nanoclay on mechanical, thermal, barrier and biodegradability have been investigated by Marques et al [37] and Park et al [38]. McGlashan and Halley [39] have prepared polyester-compatibilized starch-clay hybrid nanocomposites in melt. Plasticized cellulose – organically modified clay nanocomposites in melt – hves been produced and reported by Misra et al [40]. Synthetic biopolymers like poly (butylenes succinate), aliphatic polyesters, poly (caprolactone) etc. generally undergo melt intercalative nanocomposites formation due to processing ease. Synthesis of poly (butylenes succinate)-organically modified clay has been reported by Sinha Ray et al [41]. Lee et al have reported melt intercalative clay-based nanocomposite formation from biodegradable aliphatic polyesters [42]. Bharadwaj et al have reported simultaneous clay nanocomposite formation and crosslinking of the polyester matrix using methyl ethyl ketone peroxide [43]. Di et al have reported poly (caprolactone)-clay nanocomposite formation in melt [44].

In situ nanocomposite formation often begins with monomer(s) and nanofiller sol or solvated polymer and nanofiller precursor(s) or both monomer(s) and nanofiller precursor(s). The precursors exclusively produce first generation nanofillers. Efficient adsorption of polymers over nanofiller surfaces primarily controls filler aggregation. The following schematic (Figure 5) clearly demonstrates this.

Bandyopadhyay et al have demonstrated this using three different polymer matrices in which nanosilica has been allowed to generate under sol-gel process [45]. They have shown that the most interactive polymer matrix not only prevents silica aggregation but at the same time allows quantitatively much higher silica generation as well. Very recently, in a review article, Oh and Park described the synthesis of superparamagnetic nanocomposites via *in situ* method where the polymer is synthesized in the presence of these nanoparticles [46]. Various types of nanoparticles are in use, such as: FePt, FePd, CoPt, CoO, $CoFe_2O_4$, MnPt etc. Qu et al have reported preparation of iron oxide-chitosan nanocomposites via two-step method [47]. In the first step, iron oxide nanoparticles are synthesized from mixed precursors of iron chloride and iron sulphate, and have been surface-modified with oleic acid. In the second step, the nanoparticles are added into chitosan. Conversely, Satarkar and Hilt have reported poly (N-isopropylacrylamide)-iron oxide nanocomposites starting from monomer N-isopropylacrylamide [48]. They have polymerized the monomer in the presence of stabilized iron oxide nanoparticles. Brayner et al have synthesized monometallic (Au) and bimetallic (AuNi) nanoparticles *in situ* in aqueous gelatin to form the nanocomposites [49]. Similarly, Wang et al have synthesized trimetallic CdHgTe assembly in gelatin starting from the respective precursors [50]. Li et al have reported *in situ* silver nanoparticles containing cellulose nanocomposites [51]. Boccaccini et al have recently reviewed polymer nanocomposites synthesized with the more infrequently used bioactive nanoglass particles/fibres [52]. Bozanic et al have reported *in situ* silver nanoparticles stabilized with glycogen protein polymer [53]. Yeo et al have reported two step synthesis of silver sulphide-poly (3-hydroxybutyrate) nanocomposites [54]. The first step is the formation and stabilization of silver sulphide nanoparticles, and the second step is addition into poly (3-hydroxybutyrate) for nanocomposite synthesis. There is also evidence of nanocomposite formation with nanoclay. Paul et al have synthesized poly (lactic acid)-clay hybrid nanocomposites via coordination-insertion mechanism [55]. Messersmith and Giannelis have reported synthesis of poly (caprolacone)-clay hybrid nanocomposites via the intercalative polymerization route [56].

Characterization of Polymer Nanocomposites

Polymer nanocomposites are primarily characterized for nanofiller dispersion. Morphology offers direct evaluation of the dispersion state of different classes of nanofillers. The persuasive morphology tools are atomic force microscopy, transmission electron microscopy and field emission scanning electron microscopy. For crystalline nanoparticles such as layered silicates, hydroxides and graphene, additional information can be obtained from wide- and small-angle X-ray diffraction studies and Raman spectroscopy. Atomic force microscopy has been the most authentic morphology tool available till date to analyze nanostructures, due to the availability of several sophisticated accessories. These additional tools help to evaluate some important aspects like polymer-nanofiller interface study, bulging height of the fillers and their dispersion depth from the mean composite surface. Figure 7 exemplifies such an image of high hydrolyzed grade poly (vinyl alcohol)-nanosilica composites. The gradation of colors clearly illustrates bare silica, bare polymer and silica-polymer interfaces. The height image gives an idea of the bulging nature of the surface. For anisotropic nanofillers, width can be read from these pictures to assign the number of layers stacked together. It would confirm the nature of nanocomposites – whether it is more exfoliated or less.

Transmission electron microscopy also gives excellent morphological evidence, but sample preparation is too tedious. Nanofillers appear as dark domains inside the colorless polymer matrix. Due to huge phase contrast, the shape and size of the nanofillers are clearly read from the photomicrographs. Alternatively, image processing software can be used to calculate size distribution and fractal dimensions of the nanofillers from such photomicrographs. Like atomic force microscopic images, thickness of the anisotropic nanofilllers are calculated to categorize resultant morphology. High resolution scanning electron microscopy offers surface morphology like atomic force microscopy but with less precision. Frequently, the images are associated with beam-cracking if one tries to increase the picture resolution. Moreover, it is difficult to analyze moistened samples as the vacuum inside the instrument is lost due to evaporation of moisture on beam-heating. All of these three morphology tools scan a very small area of the sample. A picture of bigger portion of the sample can be obtained using the energy dispersive X-ray dot mapping tool associated with both transmission electron microscopy and scanning electron microscopy. This virtually does two things – firstly it scans a much larger area of the sample (if the images are taken at low magnifications), and secondly, it confirms the presence of the intended nanoparticles by identifying them in terms of elemental analysis. One such example is shown in Figure 8. It is an energy dispersive dot-mapping of dispersed silica nanoparticles inside a poly (vinyl alcohol) matrix. The magnification is only 2000X. The white dots in the black background (silicon signals) identify these as silica, and its uniform coverage over the whole scan area states that it is uniformly dispersed inside the matrix.

Small-angle X-ray diffraction analysis gives an idea of the average filler morphology. A more irregular shape with bigger size indicates gross aggregation of dispersed nanofillers inside the polymer matrices. Wide-angle X-ray diffraction together with morphological evidence provides information on the exfoliation/intercalation states of layered nanofillers such as layered silicates and hydroxides and graphenes. Similar information is also obtained from Raman spectroscopy on such crystalline fillers. Physico-mechanicals of polymer nanocomposites such as mechanical, dynamic mechanical, rheological, thermal, electrical, optical, photo acoustic etc. are essentially the manifestation of the degree of nanofiller dispersion followed by adhesion with polymer segments. Since these are more application-specific, a detailed discussion is out of scope of the present chapter.

Biological Implications of Polymer Nanocomposites

Tissue engineering and drug delivery are the prime areas that have been vividly explored so far for applications of bionanocomposites. Scaffolds for bone tissue engineering demands a combination of properties such as a) biocompatibility, which is needed to prevent inflammation or immunogenicity/cytotoxicity and b) sufficient mechanical strength to withstand free handling. As already stated, biopolymers are excellent biocompatible systems and thus often meet the first criterion but frequently fail to meet the next one [57]. Intrusion of bionanocomposites is primarily to offer greater mechanical strength. Sometimes, it may also confer porosity to the matrix, often helpful for bone tissue engineering [58]. Nevertheless, a high order of selectivity requires choosing of nanomaterials for such applications. Generally, nanodimensional hydroxyapatite and bioactive glass are the most preferred nanomaterials for such applications. Poly (lactic acid), poly (glycolic acid), poly (hydroxyalkanoates) and its blends, poly (caprolactone) etc. are examples of a few biopolymers that have been used as a matrix in tissue engineering applications so far. Peter et al have recently used chitosan- and gelatin-embedded bioactive glass nanoparticles as scaffolds in alveolar tissue engineering [59]. Han et al studied the antimicrobial activities of chitosan-nanoclay hybrid composites [24]. Gelatin-coated hydroxyapatite nanorods for similar kind of applications have been reported by Chen et al [19] and Chang et al [20]. Kotela et al have claimed a combination of poly (caprolactum), nanoclay, silica and calcium hydroxide as bone tissue substitute [58]. Boccaccini et al have recently reviewed the use of several nanobioactive glass-based biopolymers in bone tissue engineering [52]. Puppi et al have provided a list of natural and synthetic biopolymer-based composites used in bone and cartilage repair [60]. Other than bioactive glasses/hydroxyapatite nanofillers, the antimicrobial activity of several other nanofillers has also been investigated. Bo˘zani´c et al have experimented with the antimicrobial activities of silver nanopartilcle embedded in glycogen biopolymer [53]. Li et al have tested the same in cellulose matrix [51].

Conventional drug therapy, both oral and intravenous, has several disadvantages, like quick erosion of the drug inside the body, multiple administration causing drug overload, and undue side effects. Recently, polymer hydrogels have been investigated as drug carriers for controlled release, to widen the therapeutic window with a smaller quantity of drugs. This technique has been of particular interest for drugs with low half life, eg diclofenac sodium. Aguzzi et al have first reviewed the potential of bare clay in such applications, since the drug molecules strongly interact with clay due to its surface active groups [61]. Depan et al have studied the cell proliferation and controlled drug release behavior of novel chitosan-g-lactic acid containing nanoclay [62]. Chitosan-organoclay hybrid nanocomposites in controlled drug release applications have been investigated by Wang et al [63]. Liu et al have studied the effect of hydroxyapatite nanoparticles on the controlled release mechanism of chitosan [64]. Very recently, Oh and Park reviewed the potential of superparamagnetic iron oxide nanopartilces embedded in different biopolymer composites in several biomedical applications, such as contrast increase in magnetic resonance imaging, targeted drug delivery, hyperthermia, biological separation, protein immobilization and biosensors [46]. Qu et al employed iron oxide nanoparticles in chitosan for the treatment of hyperthermia [47]. The author and his research group have been actively engaged in developing polymer-nanocomposites-based hydrogel patches for controlled transdermal drug delivery. Excellent results have been obtained after nanosilica and functionalized carbon nanotube addition in these patches, and the advent of these nanomaterials has significantly increased the drug retention capability of these hydrogels.

Conclusions

An attempt has been made to provide a brief but resourceful account of synthesis, characterization and biomedical applications of important biopolymer nanocomposites pertaining to different types of nanofillers. This area is still very green, and a lot of new investigations are being carried out in different parts of the world with various results. The future challenge is to scale up these laboratory observations to the commercial arena in order to serve mankind.

References

1. http://jnm.snmjournals.org/cgi/content/full/48/7/1039/FIGI, access date: 14.03.2011.

2. S. Pavlidou, C.D. Papaspyrides. Prog. Polym. Sci. 33, 2008, 1119.

3. A. Ganguly, A. K. Bhowmick. Nanoscale Research Letters, 3, 2008, 36.

4. T. Bhunia, L. Goswami, D. Chattopadhyay, A. Bandyopadhyay. Swelling de-swelling studies after freeze thaw treatment of nano-silica reinforced poly (vinyl alcohol) based organic-inorganic hybrid hydrogel. A paper presented at International Conference on Nanoscience and Technology, Feb: 17-20, 2010, Mumbai, India.

5. F. Oliveira, S. R. Monteiro, A. Barros-Timmons, J. A. Lopes-da-Silva. Carbohydrate Polym. 82, 2010, 1219-1227.

6. A. L. Daniel-da-Silva, F. Pinto, Lopes-da-Silva, T. Trindade, B. J. Gordfellow, A. M. Gil. J. Colloid Interface Sci. 320, 2008, 575-581.

7. V. M. Guňko, P. Pissis, A. Spanoudaki, A. A. Turova, V. V. Turov. Colloids Surfaces A: Physicochem Engg. Aspects 320, 2008, 247-259.

8. A. Bandyopadhyay, M. De Sarkar, A.K. Bhowmick, J. Mater. Sci. 41, 2006, 5981.

9. J. Bandyopadhyay, S. Sinha Ray. Polymer, 51, 2010, 1437-1449.

10. J. J. George, A. Bandyopadhyay, A. K. Bhowmick. J. Appl. Polym. Sci. 108, 2008, 1603.

11. A. Gómez-Avilés, M. Darder, P. Aranda, E. Ruiz-Hitzky. Appl. Clay. Sci. 47, 2010, 203-211.

12. A. Giri, T. Bhunia, L. Goswami, S. Pal, A. Bandyopadhyay. Low shear rheological behavior of carbon nanotube modified thermosensitive carboxymethyl guar gum hydrogel. Paper No. A paper presented in International Conference on Advances in Polymer Technology, Feb. 26-27, 2010, Cochin, India.

13. N. Ogata, G. Jimenez, H. Kawai, T. Ogihara. J Polym Sci Part B: Polym Phys 35, 1997, 389.

14. G. X. Chen, G. J. Hao, T. Y. Guo, M. D. Song, B. H. Zhang. J Appl Polym Sci 93, 2004, 655.

15. H. M. Wilhelm, M. R. Sierakowski, G. P. Souza, F. Wypych. Polym Int 52, 2003, 1035.

16. J. P. Zheng, L. Ping, Y. L. Ma, K. D. Yao. J Appl Polym Sci 86, 2002, 1189.

17. S. T. Lim, Y. H. Hyun, H. J. Choi, M. S. Jhon. Chem Mater 14, 2002, 1839.

18. YuanQiao Rao. Polymer, 48, 2007, 5369-5375.

19. M. Chen, J. Tan, Y. Lian, D. Lin. Appl. Surface Sci. 254, 2008, 2730-2735.

20. M.C. Chang, C-C. Ko, W. H. Douglas. Biomaterials, 24, 2003, 2853-2862.

21. D. J. Greenland. J Colloid Sci 18, 1963, 647.

22. N. Ogata, S. Kawakage, T. Ogihara. J Appl Polym Sci 66, 1997, 573.

23. K. E. Strawhecker, E. Manias. Chem Mater 12, 2000, 2943.

24. Y-S. Han, S-H. Lee, K. H. Choi, I. Park. J. Phys. Chem. Solids. 71, 2010, 464-467.

25. T. Hibino. Appl. Clay Sci. 50, 2010, 1219-1227.

26. M-F. Chiang, T-M. Wu. Composites Sci. Technol. 70, 2010, 110-115.

27. S. Sinha Ray, K. Okamoto, K. Yamada, M. Okamoto. Nano Lett 2, 2002, 423.

28. S. Sinha Ray, K. Yamada, M. Okamoto, K. Ueda. Nano Lett 2, 2002, 1093.

29. S. Sinha Ray, P. Maiti, M. Okamoto, K. Yamada, K. Ueda. Macromolecule 35, 2002, 3104.

30. S. Sinha Ray, K. Yamada, A. Ogami, M. Okamoto, K. Ueda. Macromol Rapid Commun 23, 2002, 493.

31. M. Pluta, A. Caleski, M. Alexandre, M-A. Paul, P. Dubois. J. Appl Polym Sci 38, 2002, 1497.

32. K. Fukushima, D. Tabuani, C. Abbate, M. Arena, P. Rizzarelli. Euro. Polym. J. 47, 2011, 139-152.

33. K. Oksman, A. P. Mathew, D. Bondeson, I. Kvien. Composites Sci. Technol. 66, 2006, 2776-2784.

34. P. Maiti, C. A. Batt, E. P. Giannelis. Polym Mater Sci Eng 88, 2003, 58.

35. A. J. F. de Carvalho, A. A. S. Curvelo, J. A. M. Agnelli. Carbohydrate Polym 45, 2001, 189.

36. H. M. Park, X. Li, C. Z. Jin, C. Y. Park, W. J. Cho, C. K. Ha. Macromol Mater Eng 287, 2002, 553.

37. A. P. Marques, R. L. Reis, J. A. Hunt. Biomaterials 23, 2002, 1471.

38. H. M. Park, W. K. Lee, C. Y. Park, W. J. Cho, C. S. Ha. J. Mater Sci. 38, 2003, 909.

39. S. A. McGlashan, P. J. Halley. Polym Int 52, 2003, 1767.

40. M. Misra, H. Park, A. K. Mohanty, L. T. Drzal. Injection molded green nanocomposite materials from renewable resources. Paper presented at GPEC-2004, February 18–19, 2004.

41. S. Sinha Ray, K. Okamoto, P. Maiti, M. Okamoto. J Nanosci. Nanotech. 2, 2002, 171.

42. S. R. Lee, H. M. Park, H. L. Lim, T. Kang, X. Li, W. J. Cho. Polymer 43, 2002, 2495.

43. R. K. Bharadwaj, A. R. Mehrabi, C. Hamilton, C. Trujillo, M. F. Murga, A. Chavira. Polymer 43, 2002, 3699.

44. Y. Di, S. Iannace, E. D. Maio, L. Nicolais. J Polym Sci Part B: Polym Phys 41, 2003, 670.

45. A. Bandyopadhyay, M. De Sarkar and A. K. Bhowmick, Polymers and Polymer Composites 13 (2005) 429.

46. J-K. Oh, J-M. Park. Prog. Polym. Sci. 36, 2011, 168-189.

47. J. Qu, G. Liu, Y. Wang, R. Hong. Adv. Powder Technol. 21, 2010, 461.

48. N. S. Satarkar, J. Z. Hilt. J. Controlled Release. 130, 2008, 246.

49. R. Brayner, T. Coradin, M-J. Vanley, C. Mangeney, J. Livage. F. Fievet. Colloids and Surfaces A: Physicochem. Engg. Aspects 256, 2005, 191.

50. Y. Wang, C. Ye, L. Wu, Y. Hu. J. Pharmaceutical and Biomedical Anal. 53, 2010, 235.

51. S-M. Li, N. Jia, J-F. Zhu, M-G. Ma, F. Xu, B. Wang, R-C. Sun. Carbohydrate Polym. 83, 2011,422.

52. A. R. Boccaccini, M. Erol, W. J. Stark, D. Mohn, Z. K. Hong, J. F. Mano. Composites Sci. Technol. 70, 2010, 1764.

53. D. K. Bozanic, S. D. Brankovic, N. Bibic, A. S. Luyt, V. Djokovic. Carbohydrate Polym. 83, 2011, 883.

54. S. Y. Yeo, W. L. Tan, M. Abu Baker, J. Ismail. Polym. Degrad. Stab. 95, 2010, 1299.

55. M. A. Paul, M. Alexandre, P. Degee, C. Calberg, R. Jerome, P. Dubois. Macromol Rapid Commun. 24, 2003, 561.

56. P. B. Messersmith, E. P. Giannelis. Chem. Mater 5, 1993, 1064.

57. M. Sokolsky-Papkov, K. Agashi, A. Olaye, K. Shakesheff, A. J. Domb. Advanced Drug Delivery Reviews 59 (2007) 187–206.

58. I. Kotela, J. Podporska, E. Soltysiak, K.J. Konsztowicz, M. Blazewicz. Ceramics International 35 (2009) 2475–2480.

59. M. Peter, N. S. Binulal, S. V. Nair, N. Selva Murugan, H. Tamura, R. Jayakumar. Chem. Engg. J. 158, 2010, 353.

60. D. Puppi, F. Chiellini, A. M. Piras, E. Chiellini. Prog. Polym. Sci. 35, 2010, 403.

61. C. Aguzzi, P. Cerezo, C. Viseras, C. Caramella. Appl. Clay Sci. 36, 2007, 22.

62. D. Dipan, A. Pratheep Kumar, R. P. Singh. Acta Biomaterialia, 5, 2009, 93.

63. X. wang, Y. Du, J. Luo, B. Lin, J. F. Kennedy. Carbohydrate Polym. 69, 2007, 41.

64. T-Y. Liu, S-Y. Chen, J-H. Li, D-M. Liu. J. Controlled Release. 112, 2006, 88.

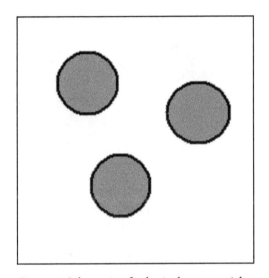

Figure 1. Schematic of spherical nanoparticles
(Resource: Bandhyopadhyay, A. unpublished data).

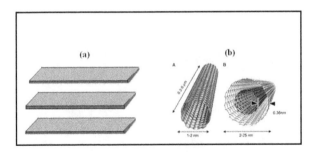

Figure 2. (a) Schematic of layered nanofillers and (b) images of single (A) and
multi-walled (B) carbon nanotubes [1]
(Resource: Bandhyopadhyay, A. unpublished data).

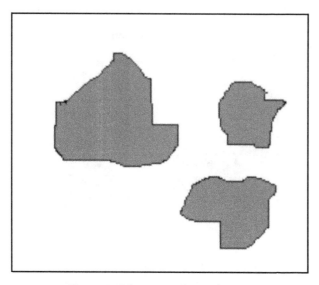

Figure 3. Schematic of nanoclusters
(Resource: Bandhyopadhyay, A. unpublished data).

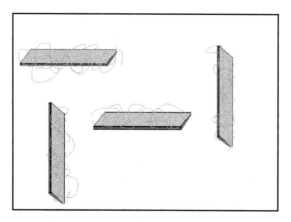

Figure 4. Completely exfoliated layered nanofillers; the polymer chains are
adsorbed over the tactoids (Resource: Bandhyopadhyay, A. unpublished data).

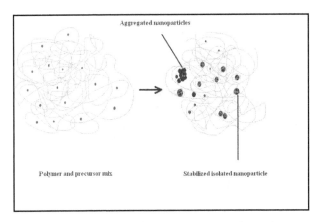

Figure 5. Schematic of the nanocomposite structure formed in in-situ polymerization processs starting from a nano-precursor; the morphology in right side shows some local aggregation of nanofillers due to insufficient polymer adsorption (Resource: Bandhyopadhyay, A. unpublished data).

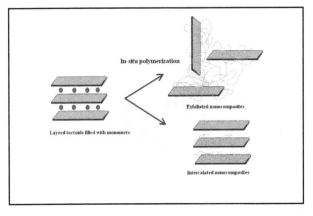

Figure 6. Schematic of intercalated and exfoliated nanocomposite formation in in-situ polymerization process (Resource: Bandhyopadhyay, A. unpublished data).

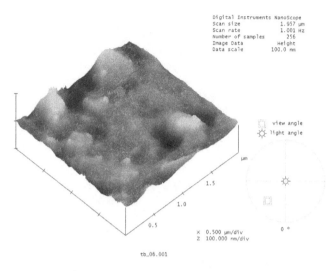

Figure 7. Atomic force microscopic height image of poly (vinyl alcohol)-nanosilica hybrids (Resource: Bandhyopadhyay, A. unpublished data).

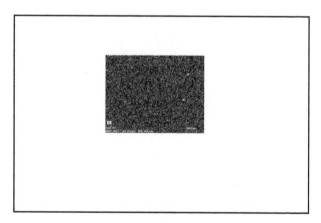

Figure 8. Energy dispersive silicon dot mapping in poly (vinyl alcohol)-nanosilica hybrid composite (Resource: Bandhyopadhyay, A. unpublished data).

INDEX

Author Biography

Arunava Goswami graduated from Tata Institute of Fundamental Research, India and did his post-doctoral studies at Harvard Medical School, USA, where he was mentored by Prof. Linda Buck, Nobel Laureate 2004. He has more than 50 international peer reviewed publications, 8 patents, 1 book and more than 80 published abstracts from national and international conferences to his credit. He is presently a Professor in the Biological Sciences Division of Indian Statistical Institute, India.

INDRANI ROY obtained her master's degree in Science from Delhi University. She is a National Science Talent (National Council of Education and Research, GoI) award holder. After a long hiatus she rejoined academics as a fellow in the Women Scientists Scheme of DST, GoI in 2007. Currently she is associated with Dr. Arunava Goswami's lab at the Indian Statistical Institute as a researcher in application of Nanoscience to agricultural situations.

Samrat Roy Choudhury is currently a Postdoctoral Research Assistant at the Purdue University, USA. Dr. Roy Choudhury has graduated from the Indian Statistical Institute, India. He has made an excellent credential in research, which is reflected in his published peer reviewed articles, patents, books and other relevant scholarly activities. He has been awarded with the 'Best Scientist Award in Biotechnology' at The 18th State Science & Technology Congress (2011) organized by the West Bengal Science & Technology Council and Department, India.

DIPANKAR CHAKRAVORTY: Prof. D Chakravorty obtained his PhD degree from Pennsylvania State University, in Solid state Technology in 1965. He is an Emeritus Professor and INSA Honorary Scientist, working in MLS Prof. Unit, Indian Association for the Cultivation of Science, Kolkata. His main research interests are chemistry and physics of nanomaterials with special emphasis on nanocomposites.

SHILPI BANERJEE: Shilpi Banerjee received her M.Sc. degree in Physics from The University of Burdwan in 2007. She is pursuing PhD programme by doing research in MLS Prof. Unit and Dept. of Materials Science, Indian Association for the Cultivation of Science as a CSIR senior research fellow. She is working on synthesis and characterization of multiferroic nanocomposite and mesoporous materials.

DHRITI RANJAN SAHA: Dhriti Ranjan Saha received his M.Sc. degree in Physics from Tripura University (Central University) in 2007. He is pursuing PhD programme by doing research in MLS Prof. Unit and Polymer Science Unit, Indian Association for the Cultivation of Science as a CSIR senior research fellow under Jadavpur University. He is working on synthesis and characterization of nanoglass composite materials.

Panchanan Pramanik obtained his doctoral degree from IIT Kharagpur in 1977 and is presently a faculty member in the Department of Chemistry, IIT Kharagpur. He is one of the leading scientists of India in nano-materials. He was on the editorial board of Journal of Nanoscience and Nanotechnology, USA. He has co-authored an encyclopedia on nano-science and nano-technology published from USA.

Arindam Pramanik, a fresh graduate student is associated with a project in the department of Biotechnology and Life Science at Jadavpur University. He obtained his Master's degree from Kalyani University in Microbiology. He was earlier associated with a project at the Chemistry department of IIT-Kharagpur.

Abhijit Bandyopadhyay is an Assistant Professor in the department of Polymer Science and Technology, University of Calcutta since November 2008. He has received several awards like: Young Scientist Award (2005) by Materials Research Society of India, Kolkata Chapter, Young Scientist Award (2009) by Department of Science and Technology, Govt. of India and Career Award for Young Teachers (2010) by All India Council for Technical Education, Govt. of India. His research interests are: Polymer Nanotechnology, Hydrogels in Drug Delivery, Reactive Blending, Polymeric Flocculant and Waste Management.

RATAN LAL BRAHMACHARY, ex professor and head, Unit of Embryology, Indian Statistical Institute was educated at the Universities of Dacca, Calcutta and Hamburg. He also worked at the Marine Biological Labs in Naples, Palermo & Banyuls-sur-Mer. His main areas of interest have been molecular embryology of invertebrates,  biological clocks and chemical signals in animals. He has published in leading journals such as International Reviews of Cytology, Advances in Morphogenesis, American Naturalist, and Nature etc. He is also a science populizer with many books and articles in both English and Bengali.

www.ingramcontent.com/pod-product-compliance
Lightning Source LLC
Jackson TN
JSHW081313130125
77033JS00002B/8